CONNECTED STRATEGY

HARVARD BUSINESS REVIEW PRESS
BOSTON, MASSACHUSETTS

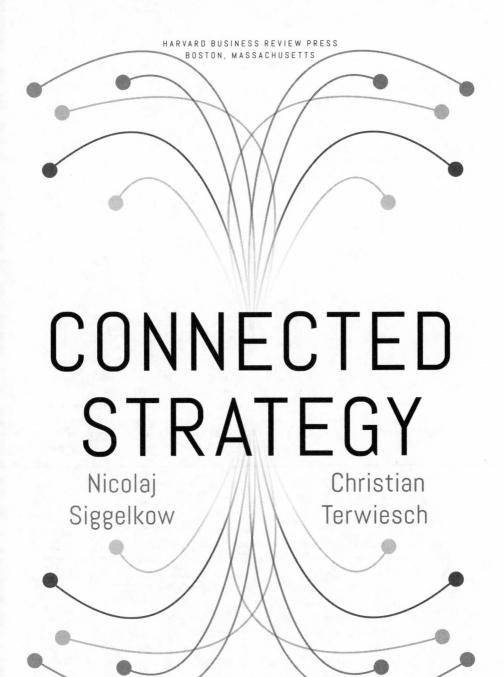

CONNECTED STRATEGY

Nicolaj
Siggelkow

Christian
Terwiesch

Building
Continuous Customer Relationships
for Competitive Advantage

Copyright 2019 Harvard Business School Publishing Corporation

Printed in the United States of America

10 9 8 7 6 5 4 3 2

The web addresses referenced in this book were live and correct at the time of the book's publication but may be subject to change.

Library of Congress Cataloging-in-Publication Data

Names: Siggelkow, Nicolaj, author. | Terwiesch, Christian, author.
Title: Connected strategy : building continuous customer relationships for competitive advantage / Nicolaj Siggelkow, Christian Terwiesch.
Description: Boston, Massachusetts : Harvard Business Review Press, 2019. | Includes bibliographical references and index.
Identifiers: LCCN 2018047850 | ISBN 9781633697003 (hardcover)
Subjects: LCSH: Relationship marketing. | Customer services—
Technological innovations. | Communication in consumer education.
Classification: LCC HF5415.55 .S55 2019 | DDC 658.8/12—dc23 LC record available at https://lccn.loc.gov/2018047850

ISBN: 978-1-63369-700-3

eISBN: 978-1-63369-701-0

The paper used in this publication meets the requirements of the American National Standard for Permanence of Paper for Publications and Documents in Libraries and Archives Z39.48-1992.

To my dad, who would have been
the proudest reader
(NS)

To my parents, who taught me more about
business than any school ever could
(CT)

CONTENTS

Prologue: The Magic of Connected Strategy   *ix*

1. The Connected Strategy Framework   1

PART ONE
THE REWARDS OF CONNECTED STRATEGIES

2. Breaking the Trade-off between Superior
    Customer Experience and Lowering Costs   21

3. Workshop 1   49
    *Using Connectivity to Provide Superior Customer
    Experiences at Lower Costs*

PART TWO
CREATING CONNECTED CUSTOMER RELATIONSHIPS

4. Recognize, Request, and Respond   63
    Building Connected Customer Experiences

5. Repeat   91
    *Building Customer Relationships to
    Create Competitive Advantage*

6. Workshop 2   121
    Building Connected Customer Relationships

PART THREE
CREATING CONNECTED DELIVERY MODELS

7. Designing Connection Architectures   147

8. Revenue Models for Connected Strategies   173

9. Technology Infrastructure for
   Connected Strategies   195

10. Workshop 3   217
    Building Your Connected Delivery Model

Epilogue: Seizing the Connected Strategy Potential   *235*

Sources   *239*
Index   *251*
About the Authors and Acknowledgments   *261*

The Magic of Connected Strategy

The practices of entertainment giant Disney illustrate a seismic shift in how firms create competitive advantage through what we term *connected strategy*. Firms with a connected strategy fundamentally change how they interact with their customers and what connections they create among the various players in their ecosystem. At its core, a connected strategy transforms traditional, episodic interactions with customers into connected customer relationships that are characterized by continuous, low-friction, and personalized interactions.

For years, Disney has organized mini-camps for children as part of its cruise operations. Stressed-out parents on a Disney cruise can drop off their kids for a few hours in order to get some private time on board. Disney has always taken its responsibility for the children in its care seriously. Until 2005, this meant stopping camp activities every thirty minutes for roll call, thus creating extra work for the staff and interrupting the fun. In 2005, Disney began using monitors to track their small campers—monitors originally developed in the medical field to keep track of dementia patients who were at risk of wandering off. The children were outfitted with little bracelets that identified them and pinpointed their location within the ship.

This was the starting point for a whole new Disney experience named MagicPlus. With the ability to know the identity and location of every guest, Disney soon asked how else it could use MagicPlus to enhance the guest experience and improve operational efficiency.

The answers that emerged touched on a remarkably broad range of its theme park operations. As part of its meet-and-greet program, Disney characters routinely interact with young guests in the park. Before MagicPlus, later renamed MagicBand, Disney's cast members posing as Mickey Mouse or Captain Jack Sparrow knew little about the children. With the MagicBand, Mickey and the captain not only knew each child's name but were also aware of the family's prior visits to Disney theme parks around the world. If six-year-old Sydney met Mickey Mouse last year in Orlando and this year was at the Anaheim park, Mickey would "remember" the first encounter, making the child's experience truly magical.

Beyond creating enhanced customer experiences, the MagicBand also improved park operations, thus reducing Disney's costs—a truly magical outcome from a business perspective. From tracking food orders to handling guest complaints, from identifying each guest to having access on a tablet to all their prior interactions with cast members, efficiency was greatly improved. Moreover, the MagicBand allowed guests to make reservations at busy attractions for predefined time windows, slashing their wait times. This enabled Disney to direct the stream of visitors and jump-start operations at the beginning of the day, increasing the overall number of visitors that could be handled by a theme park while maintaining a great visitor experience.

Connected strategies unfold in a rapidly changing environment. Given its success, one would imagine that Disney would equip every park with the MagicBand. But this is not the case. When Disney opened its Shanghai Disney Resort, it decided against the move. It wasn't that the Chinese failed to appreciate the magic. Instead, by the time of the park's opening, a more efficient alternative had arrived, and almost every visitor already had this magical technology in his or her pocket: a smartphone. Equipped with the right apps, today's phones offer all the information and access to Disney that the MagicBand had provided.

The common theme of the Disney story and many more case studies in this book is that we are living in a world where new forms of connectivity are transforming the way companies do business. Connected strategies allow you, as an executive, to create superior customer experiences while simultaneously achieving dramatic improvements in operational efficiencies. In short, connected strategies can substantially relax the trade-off that firms have traditionally faced between providing superior customer experiences and lowering costs. Adopting connected strategies allows firms to create a formidable competitive advantage. Not surprisingly, their rapid rise is creating new winners and losers in its wake.

The technologies behind connected strategies are improving at a dizzying speed. The world's estimated three billion smartphones pack the power of supercomputers from only a decade ago. The Internet of Things enables instant communication among systems that couldn't talk to each other before. Wearable health and fitness trackers now rival traditional medical devices in their accuracy. And recommendation systems driven by artificial intelligence deliver insights faster than humans ever could. With all these advancements, magical user experiences are coming to life in many industries. What is fascinating, however, is that the technology per se is usually relegated to a supporting role. The key innovation of connected strategies lies in their revamping of a firm's business model. Consider the following four examples.

Amazon redefined how retailers interact with customers. Until recently, customers were forever making long shopping lists and driving to various stores. Now, they can tell Amazon's Alexa to order food, clothing, or just about anything else, and the products are delivered to the customer's house, sometimes within hours. Beyond Alexa, Amazon also introduced the Dash Button, a small Wi-Fi device that customers can attach to the refrigerator, washing machine, or bathroom vanity and press to reorder bottled water, detergent, toilet paper, and much more.

College textbook publishing is also going through a fundamental transformation. In the old days, students would buy or rent their textbook, read the assigned chapters (or at least intend to do so), and then prepare for their final exam by tackling a set of practice problems at the end of each chapter. Now, McGraw-Hill Higher Education, for example, has abandoned the word *book* and instead aims to sell digital learning experiences. Not only are the books fully digital, they are also smart. As students go through the semester, technology tracks their reading and feeds the data back to the professor and the publisher. When a student is struggling with an assignment, the book redirects the student to the appropriate chapter and potentially offers a short video message on how to handle a similar assignment. Each student thus receives a curated and customized learning experience, rather than a standard textbook, and textbook publishers are moving beyond their traditional role to become tutors as well.

Customers used to interact with Nike once a year to buy running shoes, and that interaction was actually with a shoe retailer, not Nike. Now, customers purchase a wellness system that includes a chip embedded in the shoes, software that analyzes their latest workout, and a social network with other Nike runners for support. Customers interact daily with Nike, allowing the company to transform itself from shoe manufacturer to purveyor of health and fitness services, and to coach customers to achieve their goals.

Millions of consumers have embraced wearable technologies and let devices such as Apple Watch or FitBit track their daily lives. Some extreme users, also known as quantified selves, are measuring every aspect of their bodies, from glucose levels to body weight to nutrition to sleep cycles. When Apple knows more about a patient than her doctor, this has major implications for the health care industry. Many digital delivery systems are integrating this data stream with the patient's electronic medical record. In the old days, patients and doctors would see each other during periodic visits. Now, changes in body weight or blood pressure and medication compliance are reported to the care provider daily, prompting timely action triggered by abnormalities in the data feed. At the cutting edge, firms like Medtronic have gone one step further. Some implanted devices not only track health data and communicate it but are even smart enough to take action automatically when they detect abnormal patterns in clinically relevant variables.

As you might have noticed, there are two common threads to these developments. First, firms are fundamentally changing how they connect with their customers. Rather than having episodic interactions, firms are striving to be connected in a continuous way, providing services and products as the needs of customers arise, sometimes even before customers have become aware of their own needs. These firms create a connected relationship with their customers.

Second, firms not only address a wider range of needs, they do it at lower costs. For most of its business, Amazon doesn't need expensive retail outlets; the customized, artificial intelligence–based tutoring by McGraw-Hill forgoes expensive instructors; the motivation to achieve goals in Nike's system is created by a peer-based network, not personal trainers; and the implanted medical devices that automatically take action also save money by avoiding hospitalizations. The potential of connected strategies is to create customer experiences that feel like magic while improving operational efficiency to enhance financial success.

Given their tremendous potential, connected strategies create great opportunities for you—but also for all of your current and future competitors! Connected strategies will lead to disruption in many industries. At a time when a mobility platform is valued more than some of the biggest car companies in the world, increases in connectivity can often be seen more as a threat than an opportunity. But, in almost any industry, potentially disruptive threats come and go, and at any given moment you might fear disruption from a dozen different new ventures. Which ventures will survive? Which ones are truly disruptive? We hope that the frameworks and tools in this book will not only help you in creating your own connected strategy but also provide you with a new perspective, allowing you to separate technological hype from true strategic challenges.

What is a connected strategy? What tools and frameworks can you use to build one for your organization? How does it help you create a competitive advantage? What are great examples to learn from? Answers to these questions are at the heart of this book.

1

The Connected
Strategy Framework

A good way to start understanding connected strategy is to consider the traditional relationship between customers and companies. Traditional interactions start when customers realize they have an unmet need. This need could be the desire to see Mickey Mouse and ride a roller coaster, the dream of mastering financial accounting, or the urge to get into shape before summer arrives. Customers then figure out how they want to fulfill this need. They browse theme parks on Expedia, they look for accounting books at Barnes & Noble, or they consult with friends or the local gym on how to train for a triathlon.

At some point, customers attain a level of knowledge that sparks action to put some money on the table. They book a ticket to Disneyland, they buy an expensive textbook, or they sign up for a weeklong training camp. But there is considerable friction in the traditional transaction: customers spend a significant amount of effort to search, request, and receive the product or service they desire.

Firms sit on the other end of these traditional transactions. Yes, they can use marketing dollars to influence the customer along the journey to place an order, but they have limited connections to that customer.

Their episodic interactions start only once the customer has placed an order, and they end on delivery of the product.

In traditional interactions, firms work hard to provide high-quality products and services as quickly as possible and at a competitive cost. They manage and perfect their marketing and operations within the model of episodic sales, but they are inherently limited by the lack of deep connections with their customers. Traditional episodic interactions between customers and firms usually require customers to invest significant effort in figuring out a solution to their needs, then requesting and receiving the product or service. Moreover, there often exists a gap between what the customer wants and what the firm provides. This gap can be a temporal gap (the customer must wait), or a gap between what the customer really wants and what the firm has to offer.

A firm that is able to move from episodic interactions with its customers to a connected relationship overcomes these shortfalls. Consider again the power of the MagicBand. Disney used to have only a handful of interactions with its visitors, and those happened at well-defined intervals—when they came to the park and bought a ticket, or when they ordered cheeseburgers at the restaurant. Now, sensors track the guests via their MagicBand every step and every second. The MagicBand not only reduces the effort in ordering and receiving cheeseburgers or souvenirs, it tailors the experience by making suggestions to the visitor.

Similarly, McGraw-Hill originally interacted with a reader only when selling a book, and even that connection was delegated to a retailer, similar to the case of Nike. But today, every time the reader looks at the book or tackles a practice problem, a connection is established that allows the publisher to learn about the reader, curate its offering, and coach the student when he or she is stuck. Meanwhile, in health care, the connected strategy moves the doctor-patient relationship from an episodic encounter every few months to a continuous flow of data from the patient to the care team, enabling medical needs to be addressed before they become severe.

Moving away from episodic interactions toward a connected relationship turns a theme park into a magical experience, transforms a book publisher into a creator of learning journeys, and revamps a hospital system into a proactive care organization. Such deeply connected relationships create more loyalty and higher profits.

Connected strategies don't just happen; they need to be carefully designed. They have two key elements: a *connected customer relationship* and a *connected delivery model*. The connected customer relationship is what delights the customer. The connected delivery model is what allows the firm to create these relationships at low cost. Each connected customer relationship and delivery model is the result of strategic choices along several design dimensions. Let's look at these in figure 1-1.

At the heart of the connected strategy is the connected relationship between customers and a firm. We find it helpful to think about four design dimensions of a connected customer relationship, which we will refer to as the *four Rs* of connected relationships. First, the information that flows from the customer to the firm allows either party to *recognize* a customer need. Once a need is recognized, the customer or firm identifies a product or service that would satisfy this need, leading to a *request* for a desired option. In turn, this triggers the firm to *respond*, creating a customized, low-friction customer experience. By interacting with customers frequently, a firm is able to *repeat* the

FIGURE 1-1

The connected strategy

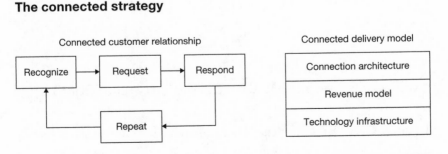

interactions with its customers, allowing it to continually refine the cycle of recognize-request-respond and to convert episodic interactions into a true relationship with its customers.

To create connected customer relationships in a cost-efficient way, a firm needs to create a *connected delivery model*. The delivery model is the result of three key strategic decisions. First, the firm has to decide whom to connect with whom in its ecosystem. What connections need to be created between and among its suppliers, its customers, and itself? We call this the *connection architecture*. Second, the firm has to decide how money will flow through this architecture, allowing it to monetize the value that results from breaking the trade-off between customer happiness and efficiency: it has to design a *revenue model*. Lastly, the firm has to make a range of technological choices that facilitate all the elements of a connected strategy. It has to decide on its *technology infrastructure*.

This book is designed to help you both understand and create connected strategies for your own organization. We have structured the book into three parts. In part I, we show in detail how connected strategies allow you to break your existing trade-off between customer happiness and efficiency. Part II helps you understand how to build connected customer relationships. Finally, in part III, we describe how to build a connected delivery model. Each part concludes with a chapter we call a workshop. In these workshops, we offer exercises that have been tested and refined with executive education audiences. These workshops will help you assess your firm's current activities and create your own connected strategy.

To provide you with a road map of what lies ahead, here is a brief preview.

Part I: The Rewards of Connected Strategies

In chapter 2, "Breaking the Trade-off between Superior Customer Experience and Lowering Costs," we discuss several case studies that illustrate how connected strategies can overcome the trade-off between customer happiness and efficiency—a trade-off that is foundational to most traditional strategic planning frameworks. By tracking guests in a theme park, selling smart books, and taking care of patients' health rather than thinking of them as appointment slots, a connected strategy creates value by breaking the existing trade-offs between the value that a customer receives and the cost that the firm incurs. The reward of a connected strategy is providing more value to the customer at a lower cost to the firm.

We explain how current innovations in the grocery retail sector, including meal-kit delivery services, augmented reality displays, and stores without checkout lines, are increasing both customer satisfaction and efficiencies.

Using a detailed case study of the ride-hailing industry, we then discuss how firms such as Uber and Lyft not only have improved the passenger experience compared with cab companies but are able to do this at much lower fulfillment costs. By connecting passengers with drivers, ride-hailing companies have created a market for driving services. Further, by allowing prices to vary depending on supply and demand, drivers are given an incentive to work when and where it is most valuable. Matching supply with demand in a dynamic world requires new forms of connectivity that extend well beyond a person jumping into the street to flag a cab or calling a grumpy dispatcher to send a taxi. Once that connectivity is put in place, resources can be used much more efficiently.

We finish the chapter by discussing how connected strategies can lead to competitive advantage and by reflecting on the importance of data privacy in the context of connected strategies.

Chapter 3, titled "Workshop 1: Using Connectivity to Provide Superior Customer Experiences at Lower Costs," contains a series of worksheets that will start you on the process of creating a connected strategy for your organization.

Part II: Creating Connected Customer Relationships

In part II of the book, we analyze in depth how you can create a connected relationship with your customers—a relationship in which episodic interactions are replaced by frequent, low-friction, and customized interactions enabled by rich data exchange.

In chapter 4, "Recognize, Request, and Respond: Building Connected Customer Experiences," we investigate the first three design dimensions of a connected relationship. The dimension of *recognize* encapsulates the information flow between the customer and firm that leads to the recognition of a customer need. We discuss various ways in which you can shape this information flow: this flow might be initiated by the customer, or it might be autonomous. Once the information reaches the firm, the firm needs to interpret and convert it (or help the customer to convert it) into a *request* for a desired option. Lastly, the firm needs to *respond* to this request and fulfill the desired option in a low-friction manner. This full interaction between customer and firm creates a connected customer experience. Through our research, we have identified four different types of connected customer experiences. These are:

Respond-to-desire

Curated offering

Coach behavior

Automatic execution

Let's take a look at each, returning to our examples from the prologue. Amazon is a great case of what we call a *respond-to-desire* connected customer experience. Once the customer expresses a need, Amazon responds rapidly and conveniently. At Disney, a key function of the

MagicBand is to create a respond-to-desire experience. When a customer wants to enter a ride, pay for a cheeseburger, or open her hotel room, a swipe with the MagicBand is all that is needed.

The McGraw-Hill textbook example illustrates *curated offering*. Having many interactions with each customer allows the firm to learn about the customer's needs. With that knowledge and trust, the customer is no longer alone in finding solutions. Here, the firm and the customer look for solutions jointly. McGraw-Hill does not just help the student figure out the corporate valuation problem on page 247, but instead detects that the student is still struggling with net present value calculations and asks him to repeat the content on page 35.

Companies that create connected strategies often create more than one connected customer experience. Returning to our Disney example, it's clear that the MagicBand does more than create a respond-to-desire experience. With the MagicBand, the customer can communicate a decision that she no longer wants to take a ride on Magic Mountain. Instead, she tells Disney (or Disney knows from past experience) that she wants to experience an action ride and a yummy meal in the next two hours. Disney then takes this information and creates a personalized itinerary. Moreover, Disney is even able to customize the experience of different rides. For instance, if a visitor has created an avatar in one of Disney's video games, this avatar will appear on the "Wanted" poster that the visitor sees during the Pirates of the Caribbean ride.

We call the third type of connected customer experience *coach behavior*. Firms like Nike try to change the behavior of customers toward what is good, smart, or healthy. Nike does not force you to go running more often, but it can offer to help you achieve your fitness and health goals. Similarly, the virtual tutor in a smart textbook says, "Jeremy, you have not yet completed the assigned readings for this week," just as a wearable device starts vibrating if its owner has not left his office chair for the last few hours.

Connected devices that let health care providers intervene even before an urgent need has arisen and implanted devices that are able to take independent actions are examples of the *automatic execution* connected customer experience. A cardiologist is consulted the moment an arrhythmia is recorded by the heart-rate monitor. A digital photo album is created and sent to the customer based on many shots taken in the theme park, all done without the customer ever noticing a camera. As with many connected strategies, these deep connections can raise privacy issues, as they sit in a gray zone between Big Brother and parental love. We will explore these issues throughout the book. And, just to be clear, we don't see this as the vision for all connected customer experiences, though most of our students would happily permit textbooks to take their exams for them . . .

While these individual customer experiences already create a lot of value, once a firm is able to repeat these interactions, it has the ability to substantially refine the customer experience over time. A firm with a connected strategy is able to transform a series of customer experiences into a connected *relationship* with its customers—a key condition for a firm to create a competitive advantage. This transformation is the topic of chapter 5, "Repeat: Building Customer Relationships to Create Competitive Advantage."

We believe that many connected customer experiences will become table stakes in the future. That is why the repeat dimension is so important. It is through this dimension—the ability of firms to learn from existing interactions in order to shape future interactions—that firms will be able to create a sustainable competitive advantage. The *repeat* dimension helps firms with two forms of learning.

First, at the level of a particular customer, a firm learns how to better match the needs of this customer with the firm's existing products and services. Disney learns that Jing seems to like ice cream more than fries and theater performances more than rides, so it is able to create a more enjoyable itinerary for her. McGraw-Hill learns that Jeremy struggles with compound-interest calculations, and is able to direct his attention to material that covers exactly that weakness. Netflix learns that Venkat likes political satire, and can make more pertinent suggestions to him of what movies he would enjoy.

Second, beyond this customer-specific learning, the firm can engage in population-level learning, allowing it to adjust its existing portfolio of products and services. Disney learns that the general demand for frozen yogurt is increasing, so it can add more stands serving frozen yogurt. McGraw-Hill learns that many students struggle with compound-interest calculations, so it refines its online module on this topic. Netflix observes that many viewers like political dramas, so it licenses additional series in this genre. Moreover, population-level learning may allow a firm to know more about its customers than any of its suppliers do, thereby enabling it to create new products and services. Having deeper customer insight allows McGraw-Hill's content producers to add new educational experiences, and Netflix to move into movie production itself.

Over time, these two levels of learning have another very important effect: firms are able to address more fundamental needs of customers. McGraw-Hill might learn that its customer wants not just to learn financial accounting but in fact to make a career on Wall Street. Nike might find out that a particular runner is interested not just in keeping fit but also in training to run a first marathon. This knowledge can lead to opportunities to create an even wider range of services and to trust relationships between firms and customers that become very hard to break by competitors. To build these trusted relationships, customer data needs to be used in transparent and secure ways, a topic we return to at the end of chapter 5.

Chapter 6, "Workshop 2: Building Connected Customer Relationships," concludes part II of the book by guiding you through a series of exercises that will assist you in creating connected customer relationships.

Part III: Creating Connected Delivery Models

Once you have an idea of the type of connected relationship you want to design for your customers, the question is how to implement this

HOW THIS BOOK CAME TO BE

At the Wharton School, both of us teach in the MBA and executive educa-tion programs, and we are codirectors of the Mack Institute for Innovation Management. In the last few years, as new and existing firms disrupted their industries by fundamentally changing how they connect with their custom-ers, we have begun to think about the principles underlying their suc-cess. At the same time, more and more managers have come to Wharton Executive Education to learn how to create opportunities for their own businesses using these principles. As a result, we set out both to help managers navigate the world of the Internet of Things, the sharing econ-omy, platform strategies, deep learning, fintech (financial technology), and so on, and to provide them with a tool kit to create connected strategies for their own organizations. Our thinking was sharpened and refined through hundreds of sessions with executives and MBA students. The stories they shared of their own experiences and challenges were essen-tial feedback as we developed the framework we now call connected

relationship in a cost-effective way. You need to create a connected delivery model. These models consist of three parts, which form the subjects of the next three chapters.

In chapter 7, "Designing Connection Architectures," we lay out dif-ferent ways in which firms can reshape the network of connections among the various players in their ecosystems. Ride-hailing firms such as Uber and Lyft have created connections between previously uncon-nected parties: individuals with cars and individuals looking for a ride. Such a configuration of the value chain has its advantages, but there are many alternatives. In the world of mobility, Daimler has deci-ded to create its own car-sharing service (Car2Go), forming a direct strategy. The book you hold in your hands (or read on the screen) is the result.

As in any research, we have stood on the shoulders of giants. We have greatly benefited from the broad-ranging research by Michael E. Porter, who has investigated the impact of the internet and new technologies on strategy for a long time. The work by Adam M. Brandenburger, Harborne W. Stuart Jr., and Barry J. Nalebuff had an important influence on our discussion of value and willingness-to-pay in chapter 2, while the customer journey we discuss in chapter 4 has its intellectual foundation in the work by Ian C. McMillan and Rita Gunther McGrath. The insightful work by Andrew McAfee and Erik Brynjolfsson on the impact of new technologies on firms and society has always been stimulating for us. Finally, we have been inspired by our friends, colleagues, and coauthors David Asch and Kevin Volpp, whose pioneering work on hovering over patients while outside the hospital has influenced our thinking on connected customer relationships. Obviously, many others have influenced our thinking and work. For readers who would like to dive deeper into the related academic and applied literature, we have compiled a detailed sources section at the end of the book.

connection between a manufacturer and customers. In contrast, Zipcar, another car-sharing operation, has connections to both car manufacturers and renters, as it has to purchase cars before it can rent them out. Finally, ride-sharing service BlaBlaCar operates a peer-to-peer network of drivers who offer each other rides whenever an empty seat is available.

When implementing a connected strategy, you need to decide how much of the customer experience your firm will generate internally and how much you will delegate to other partners in the ecosystem. Moreover, you may have to create new connections between players in your ecosystem. Chapter 7 provides guidance on these decisions. As

CONNECTED STRATEGY VERSUS PLATFORM STRATEGY

The last couple of years have witnessed an enormous success of so-called platform strategies. Connected strategies are different from platform strategies; in fact, you can create value with connected strategies without being a platform.

Platform businesses are not directly involved in serving customers by providing them with goods or services. Instead, their focus is on connecting the producers and customers of such products or services. For example, Uber does not own cars. It connects drivers and their cars with passengers who want a ride. Apple's app store primarily does not sell Apple software. It connects app developers and their software with customers who want to use it. Rather than owning real estate, Airbnb focuses on connecting folks who have empty rooms, apartments, or houses with travelers in need.

A platform really serves two sets of customers. First are those who provide the products or services that are transacted on the platform, be it app developers, drivers, or landlords. And second are those who consume those products or services, such as owners of smartphones, passengers, and travelers. Platform businesses need to appeal to both sets, which is why they are often called two-sided markets. Success is typically achieved by providing a platform for payment, trust building, and dispute resolution and by providing liquidity to the market, making sure there are enough customers to make it worth the providers' effort to join the platform and vice versa.

Though platforms crucially depend on connectivity and are an important design element of our connected strategy framework, they differ along the following important dimensions:

- Platforms are a particular type of what we call connection archi-tecture. Connected strategy, however, also involves creating a connected customer relationship. For instance, Disney has created strongly connected relationships with its customers, but most of us would not refer to Disney as a platform.

- With the success (and hype) around platforms, the term has unfortunately become rather diffuse and is applied to many different models. As we will see in chapter 5, *platform* is really an umbrella term for several different connection architectures. When Amazon, a platform company, sells through its own warehouses, billions of dollars of fixed assets in real estate, buildings, and logistics are involved. When Amazon facilitates a sale through its Marketplace platform, none of these assets is involved. These are very different business models! Similarly, while Airbnb and Facebook are both platforms, one is connecting customers to individuals who serve as suppliers (of accommodation), whereas the other connects individuals who do not engage in a business transaction. Again, these are very different connection architectures that will require, for instance, very different revenue models.

- On most platforms, the relationship between customers and providers is primarily a transactional one, and the revenue model of the platform is to take a cut from these transactions. In contrast, the ultimate goal of connected strategies is to transform episodic transactions into long-term customer relationships precisely to avoid transactional pricing.

In sum, platforms and platform strategies are related to connected strategies and also can be an element of a connected strategy. Yet connectivity allows for many other forms of value creation than the formation of two-sided markets, and it is the aim of a connected strategy to take advantage of them.

we will see, there are five common connection architectures that are used across industries:

Connected producer

Connected retailer

Connected market maker

Crowd orchestrator

Peer-to-peer network creator

In chapter 7, we will describe each of these connection architectures in detail. By recognizing these, you can make the right choices for your business.

We finish the chapter by introducing the connected strategy matrix. We have found this matrix to be a very valuable framework for systematically cataloging the various activities of your competitors in your industry, as well as an innovation tool to create new ideas for your own connected strategy.

Chapter 8, "Revenue Models for Connected Strategies," adds the second design consideration for your connected delivery model: the revenue model. Some connected strategies can rely on traditional revenue models. Disney, in its theme park division, is still making most of its money from admission tickets, food, merchandise, and fees for special experiences inside the park. Contrast this with Niantic and Nintendo, two companies that also produce amazing experiences centered on fictitious characters. With Pokémon Go, the partners have leveraged augmented reality to create a technology platform that turns any place into a virtual theme park. You can play it anywhere, anytime, and you can play it for free, together with its other sixty-five million active users. Niantic creates revenues through in-app purchases that enhance the game and through sponsorships by firms that create desirable locations for the game (e.g., Starbucks and McDonald's).

Technological advances are often what make connected strategies economically feasible. Of course, everyone would like to have personalized, on-demand services that fulfill the most fundamental needs. But how can firms offer such customized services at affordable prices? Vast improvements along many dimensions, from data gathering, data transmission, and data storage to data analysis, logistics, and manufacturing, have made connected strategies a possibility. In chapter 9, "Technology Infrastructure for Connected Strategies," we will help you sift through these advances. By coming back to the connected relationships that you have designed in part II, we can identify key technologies and systematically ask which of these technologies will enable you to further advance your connected strategy.

Finally, chapter 10, "Workshop 3: Building Your Connected Delivery Model," presents a series of exercises and tools that will allow you not only to build your own connected strategy but also to get a better understanding of what is happening around you in your industry.

The concepts just introduced allow us now to formally define a connected strategy:

> A firm's connected strategy is a set of operational and technological choices that fundamentally changes

- how the firm connects to its customers by implementing the recognize-request-respond-repeat loop, which transforms episodic interactions into continuous relationships with low friction and high degree of customization, and

- what connections the firm creates among the various players in its ecosystem through the type of connection architecture it chooses and the subsequent economic value captured through a revenue model.

BUSINESS-TO-BUSINESS
CONNECTED STRATEGIES

Connected strategies are arising in both business-to-consumer and business-to-business settings. Consider Rolls-Royce's transition from a simple seller of aircraft engines to a much broader service provider enabled by deeper connectivity. Today's aircraft engines are packed with sensors that generate gigabytes of data. This data allows Rolls-Royce to have a precise understanding at the level of individual parts within an engine. While previously parts such as a fuel pump would be replaced on a fixed schedule, now Rolls-Royce can replace pumps earlier or later depending on the state of the pump. This creates significant cost savings either by avoiding engine problems and subsequent flight delays (if a pump was previously replaced too late) or by being able to use a pump longer than normally. The ability to deliver preventative maintenance has allowed Rolls-Royce to change its revenue model from selling engines to selling flying hours, aligning incentives between itself and its customers. Moreover, when engines are decommissioned and returned, Rolls-Royce knows the exact performance profile of each part and is able to recover 50 percent of the materials to remanufacture them for use as new components, reducing manufacturing costs. Being able to aggregate data across a customer's fleet allows Rolls-Royce to gain further insights. For instance, a clean engine burns less fuel, but washing engines is costly. Using fleet-level data, Rolls-Royce is able to determine the optimum time for each engine in a fleet to get washed. Finally, by adding additional data sources such as technical logs, flight plans, forecasts, and actual weather data, Rolls-Royce is able to provide insights to its customers for how to increase fuel efficiency—for instance, through improved flight schedules. Through its connected strategy, Rolls-Royce has increased the value it delivers to its customers while at the same time increasing its own efficiency.

How to Use This Book

Probably you are reading this book not just for your entertainment but rather with the intent of reflecting on how your organization creates connections to your customers and suppliers. We get it. That's why we wrote this book neither as an academic textbook nor as a scholarly treatise. The main purpose of our writing is to help you design your own connected strategy so that you can create a competitive advantage for your firm.

At the end of each part of this book, we want to pause and help you apply all the frameworks in the form of a workshop. Each workshop consists of worksheets and guiding questions that we have road-tested with a large number of our executive education participants. Use them on your own, or even better, use them to engage your team members and other stakeholders in a structured workshop.

On the website for this book, connected-strategy.com, you can find more information on how to run your own workshop. The site has suggested timelines for the workshops, as well as downloadable templates of all exercises and group assignments. We also provide you with a number of filled-out examples for the worksheets that will be useful as you apply the tools we present in this book. Lastly, the website hosts dozens of podcasts featuring executives from a wide array of industries, from consulting to education and from banking to security services.

To illustrate our ideas and frameworks in this book, we use many examples from firms around the world. Some of these examples will hopefully spark ideas for your own organization. At the same time, we want to be clear about one thing: we certainly do not expect all the firms we mention to emerge as winners in their industries. Frankly, we would be extremely surprised if they did. The world of connected strategy is developing, and many new winners and losers will be created. We believe in the power of the ideas we present in this book, but we are not offering investment tips for particular firms.

Whether you are a startup trying to disrupt an existing industry, or an incumbent firm that wants to revitalize its strategy and defend its business, whether you deal directly with end customers, or are in a business-to-business setting, we believe connected strategies will play a pivotal role in helping you achieve competitive advantage. If you don't think about connected strategies, someone else in your industry most likely will!

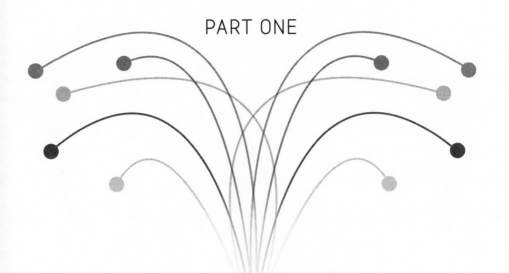

PART ONE

THE REWARDS
OF CONNECTED
STRATEGIES

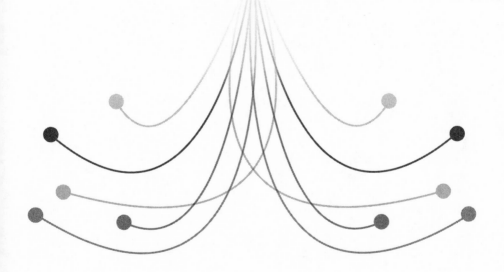

2

Breaking the Trade-off between Superior Customer Experience and Lowering Costs

Every business in every industry faces a trade-off between the quality of the customer experience and the costs of providing it. Adding a glass of champagne and extra legroom on an airplane makes for better travel, but it also increases costs. An electric car with a Tesla-style 85-kilowatt-hour battery can go faster and longer between recharging cycles, but it also costs more than the Nissan Leaf's 30-kilowatt-hour battery. A hotel providing personal concierge service leads to a superior customer experience compared with a surly desk clerk handing out local maps. In short, superior customer experiences come at the price of higher fulfillment costs. How can that be changed?

This chapter explores the fundamental promise of connected strategies: by creating deeper connections with customers and new connections between various players in an industry, firms can create new business models that redefine the existing trade-off between customer experience and cost. To illustrate our ideas, we look at case studies from grocery retail and ride hailing.

The Efficiency Frontier

To see how connected strategies completely upend the traditional cost-quality trade-off, consider supermarkets, a $600 billion industry in the United States, $500 billion in India, and more than $700 billion in China. Sam, a typical shopper, is thinking about his weekly grocery trip to buy dairy, meat, and vegetables. Traditionally, Sam considers three options: a local farmers' market, a supermarket chain such as Safeway or Tesco, and a discount market such as Aldi. What factors drive Sam's choices? Perhaps for Sam, his happiness will be most affected by the quality of the available products. For instance, at the farmers' market, the produce is organic, the meat is fresh, and the dairy comes from happy cows. Beyond the product attributes (e.g., organic vs. nonorganic), additional factors affect a customer's happiness. There are lots of ways to delight a customer, and there can be many pain points that detract from the customer experience. For instance, do customers have to drive for long to do their grocery shopping, or can they walk there? If they have to drive, how easy is it for them to find parking? How long will it take to find all the items that they want? How long will it take to pay? The list goes on.

There are many dimensions that drive how much a customer likes a product or service, including performance, features, customization, ease of access, waiting time, ease of use, and so on. We lump all of these together into a single score (think of it as a grade point average) and label this score the *willingness-to-pay* of the customer. The more you like something, the higher the maximum price you would be willing to pay for it. Economists often refer to customer utility, which captures the same concept. It is important to distinguish between the willingness-to-pay that a customer has for a particular product and the actual price that a customer pays for this product. The price for a particular product will depend not only on the customer's willingness-to-pay for this product but also on the willingness-to-pay that competitors create for *their* products and the prices they charge.

FIGURE 2-1

Efficiency frontier

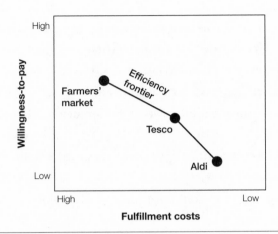

To become more attractive to customers, firms want to increase their customers' willingness-to-pay. But there is a countervailing force: the cost of creating and fulfilling such a customer experience, which we refer to as *fulfillment costs*. The better the quality of the product, or the more conveniently located the location, the higher the overall cost of fulfilling this customer experience. We can visualize this trade-off by plotting firms on a graph that has the customer's willingness-to-pay on the vertical axis and the fulfillment costs for the firm on the horizontal axis. In figure 2-1, we have plotted Sam's three grocery options. The farmers' market lives in the upper left of the graph. It creates high willingness-to-pay for customers, but its fulfillment costs are also very high, since small farms do not have scale economies in production or distribution. Costs drop as we move from left to right on the cost axis. As a result, the discount market is on the lower right. Costs are low, but willingness-to-pay for this option is also low. Supermarkets exist in between the extremes.

When we connect the dots representing the firms that are farthest out on this graph, we get the current *efficiency frontier* for this industry. This line represents the frontier because firms on this line have maximized the willingness-to-pay they have created for customers for a given cost level. Conversely, for a given level of willingness-to-pay, these firms have minimized the costs at which they can create this willingness-to-pay. Firms that are not on the frontier are at a severe disadvantage. They face competitors who can either create a higher willingness-to-pay while incurring the same cost or create the same willingness-to-pay at a lower cost. In either case, the competitor is able to provide the customer with a better deal: either a preferable product at the same price or the same product at a lower price. The efficiency frontier also illustrates that firms face a trade-off: once the firm has reached the existing frontier, higher willingness-to-pay will come at higher cost; or, conversely, cost reductions will lead to lower willingness-to-pay.

Pushing Out the Efficiency Frontier

Despite countless cooking shows on television, consumers in most developed nations have lost their appetite for spending a lot of time on meal preparation. This has led to the emergence of an entirely new product category, enabled by the internet and low transportation costs: kits that contain the ingredients for meals in customized portion sizes, delivered right to the doorsteps of subscribers. In the United States, Blue Apron initially dominated this market segment, leading to an almost $2 billion valuation in its IPO in 2017. Launched first in Germany, HelloFresh has since surpassed Blue Apron in this category. Most recently, supermarket chains such as Walmart and Albertsons have joined the party. And, you probably guessed it, wherever there is money to be made online, Amazon is not far behind—the company now sells meal kits through Amazon Fresh.

As with any case study that you will see in this book, we are not endorsing particular companies, and you should certainly not use our work to pick the next stocks you buy. Blue Apron has struggled financially since its IPO, and Uber, to be discussed later in this chapter, has had its own share of legal and financial challenges. Whatever might happen to these companies, we believe that subscription-based meal-kit deliveries and ride hailing are here to stay. The fact that both Blue Apron and Uber are facing fierce competition is tough for them as individual companies, but it demonstrates the vibrancy of the newly created market segments and thus underlines the potency of their connected business models. We will speak more about connected strategy and competitive advantage toward the end of the chapter.

How do companies such as Blue Apron operate? Customers who sign up for a Blue Apron subscription are asked to specify their preferences. From then on, every week, Blue Apron sends the customer a box containing ingredients and recipes. Blue Apron sources sustainable produce directly from smaller, often family-owned farms. All of the meats and seafood that Blue Apron provides are free of hormones and antibiotics. Quite often, Blue Apron will also include ingredients that are not very well known, such as fairy tale eggplants or pink lemons. Blue Apron's agroecologists advise farmers on crop rotation, planting dates, plant spacing, and pest management in order to enable them to cultivate unusual varieties of produce, yielding richer harvests and sustainable farming.

Why do so many customers sign up for this service? On the product side, its high-quality and sustainably sourced products put it on a relatively even level with products from a farmers' market. In addition, Blue Apron raised customers' willingness-to-pay by reducing a number of pain points along the customer journey. There is no need to spend time looking for recipes, to make a shopping list, to drive to several stores to find all the unique items, or to wait in line to pay. There are no leftover ingredients rotting in the fridge. On top of these advantages, customers learn how to cook unique meals with novel ingredients that they might not have used otherwise and might have had a hard time finding in stores. In sum, for many customers, the willingness-to-pay has increased.

But how about fulfillment costs? Despite its convenience, Blue Apron has lower fulfillment costs than a traditional farmers' market. Part of this cost advantage results from its massive scale, especially when compared with traditional farming co-ops. For instance, for some ingredients, such as the fairy tale eggplant, Blue Apron purchases nearly the entire commercially available supply.

As far as retail operations are concerned, the subscription model of Blue Apron allows the company to forecast demand for ingredients with a high degree of accuracy. This reduces Blue Apron's excess inventory. And Blue Apron does not have to worry about stock-outs, one of the major operational hassles for supermarkets. If avocados are scarce, the company can change the recipes of the week and customers will eat asparagus instead.

The forecasts also let the company help farmers manage their own businesses efficiently. Blue Apron often buys the entire crop of a farmer, providing a more predictable income stream than would be possible if the farmer attempted to sell a product at farmers' markets, where demand varies. Lastly, Blue Apron eliminates two links of the supply chain—transportation from warehouses to grocery stores, and the grocery store retail sites themselves—thereby removing real estate, utility, insurance, and other costs. By creating a curated offering, Blue Apron has pushed up the willingness-to-pay of its customers. At the same time, by creating new connections between farmers and customers—a connection architecture we will call a connected retailer in chapter 7—Blue Apron has also been able to reduce its fulfillment costs. In sum, Blue Apron has pushed out the efficiency frontier, as we illustrate in figure 2-2. The dotted line in this figure corresponds to the increased competitive advantage that Blue Apron has gained relative to farmers' markets due to its connected strategy.

Other connected strategies have arisen in the grocery space, creating an entirely new frontier in this industry. Consider the same-day delivery service Instacart. Instacart has focused on one particular pain point, the shopping itself. Instacart creates new connections between customers who need groceries and individuals who work as shoppers. It has created a connection architecture we will call a crowd orchestrator. Using the Instacart app, customers can shop virtually at a range of local stores, including grocery stores, pet supply stores, and pharmacies.

FIGURE 2-2

Pushing out the efficiency frontier

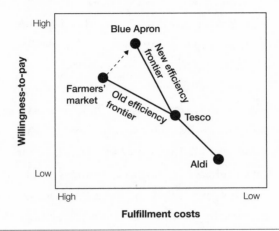

Fulfillment costs

Paid shoppers will then go to these stores and pick up the items the customer ordered. Deliveries can happen in as short as an hour or can be scheduled for later. This service raises a customer's willingness-to-pay by saving him or her time. Given that shoppers can buy for more than one person at a time, the purchases are done more efficiently, driving down cost. As we depict in figure 2-3, the main result of Instacart is also a shift of the frontier. In this case, it pushes up the willingness-to-pay more than it reduces fulfillment costs.

Around the world, we are observing a range of experiments that use connected strategies to raise the willingness-to-pay by improving the customer experience and simultaneously reducing fulfillment costs. In India, a number of e-grocers, such as BigBasket, have sprung up that allow customers to order from over twenty thousand products, including fresh fruits and vegetables, via an app and have them delivered to their homes. For a smaller group of essential items, ninety-minute delivery is also available. Customers' willingness-to-pay is increased, as the shopping can be done more quickly, while BigBasket does not have to invest in costly brick-and-mortar stores.

In South Korea, Tesco, a grocery retailer based in the United Kingdom but with extensive global operations, experimented with novel ways to engage customers in new environments. Tesco faced the challenge of trying to market itself to young urban professionals who did not have time during their day to visit stores. Tesco decided to utilize the country's high degree of smartphone adoption to reach these customers in places they do frequent: subway stations. Tesco created virtual stores on the walls of subway stations using life-size posters depicting grocery store shelves stocked with products. Customers felt as if they were standing in front of real shelves. The products featured in the posters displayed QR codes to be scanned by subway patrons using Tesco's Homeplus app. Delivery would often be scheduled for the same day so that purchases arrived as customers got home from work.

With its Homeplus app, Tesco created a respond-to-desire relationship, making it easier to shop for and receive merchandise. Moreover, their product posters also served as a coach behavior platform, reminding customers to buy groceries at an otherwise idle time, during their wait for a subway. By easing the customer journey, Tesco raised the willingness-to-pay of time-pressed customers relative to traditional grocery shopping. At the same time, it increased its sales while avoiding the cost of additional storefront real estate. Putting it all together, we can see that Tesco Homeplus effectively pushed out the efficiency frontier. It raised the willingness-to-pay and simultaneously decreased costs (see figure 2-3).

In China, Alibaba operates an increasing number of Hema supermarkets. Using the Hema app, customers can buy fresh food from home, have it prepared by chefs, and then have it delivered to their homes—all in thirty minutes. For other grocery orders, in-store shoppers fill the customer's order, putting the shopping bags on a conveyor belt that carries them to a delivery center adjacent to the store. If a customer prefers to pick his or her own food, especially fresh seafood, the customer can come and select it. Every item has a scannable barcode yielding price and product information, including the origin and even the backstory on the item. Customers use Alipay to pay at the checkout stand. Over time, Hema's app learns the customer's preference and creates more customized offerings.

FIGURE 2-3

New efficiency frontier in the grocery industry

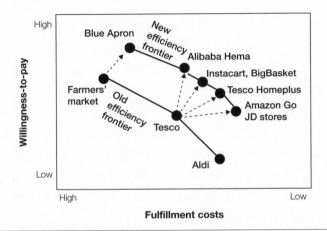

Other shopping innovations include stores that have no checkout lines at all. Instead, cameras follow customers and automatically tally up the products that they remove from shelves. Both Amazon, with its Amazon Go stores, and JD in China are pursuing this. Customers' willingness-to-pay is raised through reduced checkout hassles, while costs are reduced by lowering labor costs.

As our shopping trip around the globe reveals, grocers face a trade-off between customer delight, as captured in their willingness-to-pay, and fulfillment costs. Successful retailers should not accept the existing trade-off; they should break it.

The New Efficiency Frontier in Mobility

To see in more detail how connected strategies can relax the traditional trade-off between cost and quality, let us move from the world of grocery retailing to the world of mobility and take a deep dive into the ride-hailing industry. From Didi in China, Careem in the Middle East, Ola in India, Grab in Malaysia, and Yandex in Russia, to Uber

PARETO DOMINANCE AND SHIFTING THE EFFICIENCY FRONTIER

A strategy is said to Pareto dominate another if it can achieve both higher willingness-to-pay and lower costs. Thus, Pareto dominance is at the heart of our definition of the efficiency frontier. Every airline executive likes airplanes that provide passengers with big seats and that are inexpensive to operate. Those airplanes—economists also call them "production technologies"—Pareto dominate the ones with small seats and high operating expenses. Typically, however, airline executives have to choose between airplanes that have large seats and high costs and those that have small seats and low costs. If those are the only options, neither Pareto dominates the other and so both of them can be on the efficiency frontier. Which production technologies to choose from the efficiency frontier can be difficult to determine and will depend on market segmentation and firm strategy. (See the sources section for information on the related role of Pareto dominance and the efficient frontier in finance.)

The promise of connected strategy is that management does not have to limit its attention to making choices from among technologies that define the current efficiency frontier. Instead, we argue that by forming connected customer relationships and by using connectivity to create new connection architectures and new revenue models, it is possible to change the efficiency frontier. Consequently, successful implementations of connected strategy create Pareto-dominating strategic options—happy customers at lower fulfillment costs.

and Lyft in the United States, these companies have fundamentally changed how people think about getting around town.

The reason why ride-hailing services have gained so much popularity is that they most often provide a better service to their customers than traditional taxis and do so at lower costs. Ride hailing has shifted the efficiency frontier in the mobility industry.

To see how ride-hailing services have pushed out the efficiency frontier, we can ask the following two distinct questions that go to the heart of connected strategies:

1. *How can we change the way we connect with our customers?* How can we deepen the connections between the company, drivers, and customers? The parties being connected in this case remain the same, but the depth of their connections is increased, leading to lower costs and a superior customer experience. In essence, we put bigger information pipes between the existing entities.

2. *Can we change whom we connect?* A typical cab company has a fleet of some fifty to five hundred vehicles to meet demand. Ride-hailing services, in contrast, have many thousands of privately owned vehicles available in metropolitan areas. Moreover, they can rely on crowdsourcing to ensure the quality of their drivers, as opposed to relying on government regulators to take on the task of vetting them. This second case is one of connecting entities that previously were not connected—that is, putting in new information pipes.

Creating Deeper Connections among Existing Parties

To understand how deeper connections among existing parties can create more value, let's summarize the pain points from the perspective of the customer and the inefficiencies from the perspective of the drivers. In chapter 4, we will more formally introduce the concept of a customer journey and how to spot such pain points. For now, let's just see what a customer prefers about a ride-hailing company compared to a taxi cab—that is, how ride-hailing companies have pushed up the willingness-to-pay of customers:

Convenient ordering: Using an app is far easier than hailing a cab from a street corner, calling a dispatcher, or lining up at a taxi stand. All of these require a fair bit of customer effort; hence, none of these modes is particularly customer friendly. This is especially true at the busiest hours for cabs. For example, in New York City, 70 percent of the cabs are utilized on Thursday evenings.

Convenient payment: With cabs, each transaction begins with the pickup and ends with the drop-off. No payment information is stored beyond this. Ride-hailing companies do not require the customer to carry cash or to deal with (possibly broken) credit card machines within cabs, and payments, including tipping, happen within the app.

What about the driver? Cab drivers face several operational inefficiencies that add to the fulfillment costs without increasing the customer's willingness-to-pay. The substantial capital outlay for a cab medallion means that cabs must be in nearly constant use to be profitable. (Even after the large price drop for a medallion caused by ride-hailing companies, each taxi that you see in New York has almost as much capital tied up in it as a Rolls-Royce Phantom!) If run well, ride-hailing companies can achieve significant advantages in utilization rates. For instance, in New York, Uber's utilization rates exceed cab utilization rates by 5 percentage points. In other metropolitan areas, Uber is able to achieve more than a 20 percentage point advantage in utilization, which reflects three inefficiencies of cabs:

Finding a fare: Many cab drivers spend a good chunk of their time waiting in line at taxi stands for a passenger. For instance,

New York cabbies spend 50 percent of their time on the road without a passenger.

Routing: Even worse for the cab's efficiency is the situation in which the driver picks up a fare from a prespecified location. The meter starts running only when the customer enters the car. Human dispatchers are not only expensive, they are also limited in their computational power. Finding the optimal vehicle for a customer request requires smart algorithms and connections to the entire fleet in the market so that the best-positioned one can be chosen to fill a request—something that the cab company lacks and that leads to the lower utilization rate just mentioned.

Payment: At the end of a ride, the passenger pays the driver. This might be the most rewarding part of the ride for the driver, but is a waste of capacity. Cabs make money from driving, not from standing still while waiting to process credit card payments.

Ride-hailing companies' success in addressing these customer pain points and overcoming these inefficiencies is the first reason why they have been able to provide a superior service at lower costs. This brings us to the first question of a connected strategy: How can we deepen the connections between existing parties and get more information through these pipes?

For cab companies, this question has been answered by a number of apps, including GetTaxi, Curb, EasyTaxi, MyTaxi, and many more. These apps aim to improve the connection between a passenger and a cab; in fact, many of them were launched well before ride-hailing companies entered the market. Enabling customers to order a cab via the app turns every block in town into a virtual taxi stand. Also, payment can be streamlined as customer and company are connected through the download of an app, improving both the customer experience and the productivity of the driver in the form of higher utilization. Finally, a fleet management system can be introduced to connect all cabs in a fleet, enabling better matching between an incoming customer request and a driver, which reduces the amount of idle time, further improving driver productivity and utilization.

This improvement in connectivity is shown in table 2-1. We can think of the right-hand column in the table as a connected cab company. This company uses an app to better link to its customers and process payments, and it tracks its vehicles using GPS.

The use of technology to create deeper connections with customers can also be seen in many other examples we discussed previously:

Using the MagicBand makes food orders at Disney easier for visitors, leading to a higher willingness-to-pay, and the entire ordering and payment process is fully automated, leading to lower fulfillment costs.

Smart and fully digital textbooks are easier to produce than traditional paper books. Their embedded software also makes it easier for the school or university to grade homework assignments and exams, increasing faculty productivity.

Many health care systems now (finally) provide something that has long been a standard feature in most other industries: on-

TABLE 2-1

Benefits of creating deeper connections

Connection	Status quo cab	"Connected cab"
Passenger to vehicle	Flagging, dispatcher, taxi stand	App
Payment	Cash, credit card	Automated in-app purchase
Vehicle to vehicle	Dispatcher, two-way radio	GPS tracking, fleet management system
Improvements		• More convenient ordering • No wasted time for payment • Better routing

line booking of an appointment. Rather than going through the annoying and complex scheduling over the phone, patients can conveniently book appointments online. If more urgent care is needed, some hospitals now enable patients to videoconference with their care team directly.

Deepening the connections between the existing players can transform a business. Yet for all the enhancements implemented by the connected cab company, it is still constrained by its existing connection architecture. Let's now look at the second key element of a connected strategy: creating new connections.

Creating New Connections

The success of ride-hailing companies is not a result of being a better and more connected cab company. Their success is mostly driven by their ability to connect previously unconnected parties. Most cities have several cab companies, each with its own fleet. Ride-hailing companies, in contrast, do not own and operate vehicles. Instead, they connect individually owned vehicles throughout town to form a virtual fleet. They merely act as orchestrators, something that we discuss further in chapter 7. This virtual fleet is much larger than any cab company could ever be. This makes it easier for the ride-hailing company to cover the entire town, which increases the likelihood that a vehicle is close to a customer in need of a ride, resulting in shorter customer wait times and the higher utilizations mentioned before. Utilization is further increased because drivers don't waste time picking up and dropping off vehicles at the beginning and end of their shifts, since they simply use their own car.

Ride hailing and vehicle utilization are also improved through the practice of surge pricing. Not only are all active drivers for a ride-hailing company connected, but so are those who are currently inactive, doing other things. Ride-hailing companies often institute surge pricing in the evenings, but only at particular times. It often happens in the early evening (when folks head out), less frequently around eight o'clock at night (when folks are out), and then peaks at eleven o'clock at night (when folks want to get home). As passengers, we find surge pricing a nuisance. But it is an important component of ride hailing, just as revenue management (aka dynamic pricing) is part of selling airline tickets and hotel rooms.

Adjusting pricing in real time has two benefits. We are familiar with the first one from the airlines. Management consultants and other business executives rarely fly on Saturday afternoon. But the airlines have fixed capacity and would waste a lot of money if their planes sat idle on Saturday afternoons, so they offer steep discounts. The same happens in reverse on Monday morning when the partner in the consulting company really needs to meet that CEO in Cleveland and is willing to pay double the price. As a result, our consultant flies on Monday, and the college student heading back to Ohio State travels on Saturday. Markets are more efficient in the coordination of resources if they let prices adjust. Because the regulated cab market does not allow for price adjustments, cab owners cannot imitate this strategy. Many ride-hailing companies, however, have fully embraced dynamic pricing.

In the case of ride hailing, there is another benefit of dynamic pricing. An airline has a fixed capacity. It would love to have extra planes and pilots on Monday morning, but the costs of that capacity are too high, so there is no flexibility. Ride-hailing companies, however, have flexible supply. For the typical twenty dollars per hour, a ride-hailing driver might prefer to take the evening off. But for forty dollars per hour, during surge pricing, things are different. Previously inactive drivers get into their cars and provide the ride-hailing company and its passengers with the capacity exactly at the time they are most needed.

A remarkable aspect of surge pricing is that it not only reacts to real-time changes (when it rains, more passengers need a ride), but it can also be very targeted toward specific locations. This further helps activate and direct capacity to where it is needed the most. Surge pricing puts the right incentives in place in order for the drivers to do what is best for the overall system. When demand is low, drivers switch to other tasks, including other work and personal time off. As a result, not too many drivers are sitting idly in their vehicles and waiting for passengers, which avoids wasting labor costs. When demand picks up again, drivers provide the needed capacity and start driving. This activates capacity when needed, which avoids long waiting times for passengers.

Besides restricting entry to the taxi industry, cab medallions have traditionally served a secondary role. They provided customers with at least a modicum of trust in the car and driver. Historically, it was the role of the municipality to create trust by requiring cab operators to acquire a medallion. How are the quality and the safety of the vehicle in the ride-hailing industry ensured now that (almost) anyone can be a driver? Again, creating new connections has a role in this. Because customers rate their ride through the app, ride-hailing companies connect passengers to each other, allowing them to share their experiences. This is a much cheaper mechanism of certification than the city authority selling high-priced medallions. Driver reputation and trustworthiness are crowdsourced. Similarly, ride-hailing drivers also use the app to rate their passengers, which can help protect other drivers from picking up potentially unpleasant fares—a benefit not available to traditional cab drivers.

Building a huge virtual fleet of privately owned vehicles, adjusting fleet size to demand through surge pricing, and replacing the expensive medallions with a mechanism for crowdsourcing reputation all require a delivery model that is built on creating new connections. Such a connected strategy simply cannot be imitated by the cab companies because no company using medallions can start employing just anyone with a car who wants to be a driver.

Table 2-2 summarizes this approach of connecting previously unconnected entities in an industry. Again, this part of connected strategy—putting in new information pipes—is by no means limited to transportation:

TABLE 2-2

Benefits of creating new connections

Connection	Status quo	Ride hailing
Passenger to vehicle	Flagging, dispatcher, taxi stand	App
Payment	Cash, credit card	Automated in-app purchase
Vehicle to vehicle	Dispatcher, two-way radio	GPS tracking, fleet management system
Vehicle to vehicle outside the fleet	—	Connects all vehicles on the platform
Idle drivers and unused vehicles	—	Connects to drivers who are presently not driving
Passenger to passenger	—	Crowdsource driver reputations
Driver to driver	—	Passenger ratings leading to driver safety
Improvements	—	• Higher vehicle utilization • No medallion needed • No wasted time for driver to pick up and drop off car at the beginning and end of shift

Airbnb connects travelers to empty rooms, houses, and the people who want to play host part time or full time. This idea really goes back to prior sites such as HomeAway (which includes VRBO, Vacation Rental by Owners) that leveraged empty vacation homes (owners of vacation homes are not in their vacation homes all year). Importantly, this capacity is inexpensive because it would otherwise be unused.

ZocDoc allows physicians to post open appointment slots online, where they can be booked by patients. When an appointment is made, the patient is pleased to have found care, while the physician is happy to bring in an extra patient.

In the same way OpenTable matches restaurant tables with eaters, Kayak matches travel capacity in the form of airline seats

RELIEVING HUNGER THROUGH CREATING NEW CONNECTIONS

In the United States, seventy-two billion pounds of perfectly fine food from restaurants, catering, and event companies end up in landfills every year. At the same time, one in seven people goes to bed hungry. A lack of connectivity has created this paradoxical state. Goodr, an Atlanta-based food-waste management company, tries to alleviate this situation by creating new connections. Through an app, clients notify Goodr that they have surplus food. Goodr picks up, packages, and transports the food to nonprofits that distribute it to those in need. Goodr keeps track of the amount of food donated, making it easier for firms to take advantage of tax benefits. All parts of the transaction are logged via a blockchain application, creating a reliable record from donation to distribution of the food. Goodr finances itself through fees based on pickup volume that are less than the tax benefits that firms receive, creating a win-win situation for both firms and once-hungry clients. In the first fifteen months of operation, Goodr has been able to redirect nearly one million pounds of food—or about 850,000 meals.

with those who want to book a flight, and StubHub helps fans find tickets for games and concerts while also increasing the occupancy of the stadiums.

In sum, similar to innovators in the grocery space, ride-hailing companies have effectively shifted out the efficiency frontier: the superior customer experience comes with *lower* costs. This is what has made ride-hailing companies such a game changer in the transportation industry in many markets around the world.

Further Shifting the Frontier by Creating New Connections

How can we reduce costs further still? As a passenger waits for a ride from A to B, chances are that there are many cars driving almost the exact same route. Most of them are likely to only have one person in the car. In the United States, the average number of passengers in a vehicle is about 1.5. Why add another car (and pay for the car and the driver)? Why not just help some of these drivers who own their vehicles and are already driving with the aim of getting from A to B make some quick cash? This is the idea of BlaBlaCar.

BlaBlaCar is a European carpooling company in the business of connecting potential passengers with empty seats. Carpooling is an old idea that has been used for decades by parents, commuters, and college students. But, thanks to improved connections, carpooling has seen tremendous growth. The driver of the car is driving the distance no matter what, so the cost of adding a passenger is extremely low.

Instead of paying for labor and ownership of the vehicle, all that needs to be paid for is the gas. If the driver shares this expense with one or even multiple passengers, costs come down even more. BlaBlaCar leaves pricing at the discretion of the driver, but common price points are between ten and twenty-five cents per mile, which is about a tenfold improvement over traditional ride-hailing and even more compared to cabs. Technically speaking, the fulfillment costs are limited to the extra fuel consumption that results from adding a passenger and the added time for the extra pickup and drop-off.

Note that BlaBlaCar is doing much more than connecting potential passengers with drivers. Instead of ride-hailing companies' virtual fleets of drivers acting as service providers, *every* vehicle on the street can now be seen as a potential service provider. In this business, the lines between passengers and drivers have become blurred. In chapter 7, we will discuss such new business models based on connecting individual customers to each other as peer-to-peer networks. The following examples show the extensive reach of these connection architectures:

> What BlaBlaCar is to Uber, Couchsurfing is to Airbnb. Today, Mike sleeps in your apartment, and tomorrow, it might be the other way around.

> PatientsLikeMe connects patients to others who have similar conditions, facilitating information exchange regarding treatment options and outcomes and forming a powerful network. Initially focused on chronic conditions, such as ALS and lupus, the company has expanded to accept any patient with any condition, currently serving more than six hundred thousand members. Patients are able to learn and improve their outcomes based on previous experience for free, while researchers can gather data about what is working and develop better treatments.

> Online dating platforms such as eHarmony and Match.com are now the starting point for an estimated 5 percent of all new marriages in the United States, not to mention that well over one million babies have been born because their parents were matched by a computer algorithm. When it comes to loving relationships, both partners are, pardon the wording, customers and suppliers. Both request and respond. So, having two lonely hearts sitting at home and wishing for a partner is a waste.

To be fair, unless you are super social and enjoy being with other people rather than playing around with your phone, the willingness-to-pay for a BlaBlaCar is likely to be lower than for a ride-sharing car. Beyond the potentially chatty company in the car, finding a vehicle that gets you where you want, when you want may also require some compromise in destination or travel schedule.

Connected Strategy and Competitive Advantage

In this chapter, we saw how the grocery and ride-hailing industries are being transformed by firms using connected strategies. In both settings, we saw how connected strategies can lead to more convenience: Having the right groceries delivered to your doorstep or using the walls of a subway station as a virtual supermarket is making grocery shopping more convenient. Ordering a car via an app with automatic payment processing is likewise increasing convenience.

But connected strategies are not only about convenience and the resulting higher willingness-to-pay. Unless we provide products and services efficiently, we might not be shifting the efficiency frontier. Here is where a deep dive into the operations that create and deliver the products or services is required. In both grocery retailing and ride hailing, we saw that costs are driven by many activities, some of which might not be adding value to the customer experience. Having a large store might look nice, but if what the customer wants is simply a visual display of groceries, any wall coupled with augmented reality will do the job. And why pay a fortune for a cab medallion if what customers really want is trust, which can be produced through crowdsourcing at a much lower cost?

The efficiency frontier will be our guiding compass throughout the remaining chapters. If a firm is able to shift the frontier—that is, if a firm is able to widen the gap it creates between the willingness-to-pay of its customers and the fulfillment costs it incurs—it has taken an important step toward creating a competitive advantage.

But, unfortunately, shifting the efficiency frontier does not always guarantee that you will achieve a competitive advantage that is sustainable at least for a few years. Why not? While it is easy to see how Blue Apron has shifted the frontier and might win out over having to shop at farmers' markets, it is much more difficult to figure out why Blue

Apron's willingness-to-pay/cost gap is larger than that of HelloFresh or meal kits that Amazon or Walmart offers. As a matter of fact, Blue Apron has been struggling to retain its customers because these competitors were able to occupy a very similar location on the new, pushed-out efficiency frontier.

Likewise, it is much harder to see how Uber's willingness-to-pay/cost gap is larger than that of other ride-hailing companies—a problem that Uber was not able to overcome in China, where they lost against Didi. Just creating connected experiences is often not enough to achieve a sustainable competitive advantage. Once you've shown the world a new trick, other firms will imitate you. In order to create a sustainable competitive advantage, not only do you need to create connected experiences, you also need to create connected relationships—the heart of a connected strategy.

As we will see in chapter 5, it is especially through the repeat dimension of connected strategies that you can build a sustainable competitive advantage for your firm. Through repeated interactions, you are able to continuously refine your ability to recognize the needs of your customers, to translate those needs into a request for an optimal solution, and to have the ability to respond to these requests. Powerful positive feedback effects allow you to create long-lasting relationships with your customers and such scale economies in your connected delivery model that competitors will have a very hard time offering a better value proposition to customers.

We would like to emphasize one more point: we are convinced that *not* creating a connected strategy is a road to eventual extinction for most firms. Technological and innovative forces all point toward increased connectivity. At the same time, customer expectations are moving toward increased personalization created by deeper connections. As mentioned before, increased connectivity will become table stakes in many industries. As a result, not providing customers with connected relationships will lead to significant competitive disadvantages for your organization.

Connected Strategy and Privacy

Connected strategies are fundamentally based on a rich information flow between the customer and the firm. It is this information that allows a firm to personalize the customer relationship and to gain efficiencies in its delivery model. At the same time, customers—be they firms or individuals—are naturally wary about sharing this information. As a result, trust and privacy concerns are central to creating long-lasting connected strategies.

In the right hands, previously privately held information can be used to create value-enhancing transactions—for example, by being able to design products that better fulfill customer needs. But in the wrong hands, this information can be very harmful to customers. We find it helpful to distinguish among three types of costs that customers might incur:

1. There typically exists some personal information in our lives that we would prefer not to share with others. This might include our financial situation, medical information, our sexual preferences, our political views, or our grades in college. Even though our life might not change a lot if our neighbor knew that we only got a C– in accounting, we would prefer to keep this information to ourselves. If this data can be used by firms, individuals, or governments to harass or persecute us, or to target us in order to influence our behavior with misleading information, the damage can be tremendous. Thus, there exists a potential cost of reducing personal (emotional) safety when data is used beyond the purposes for which it was originally sanctioned by the customer.

2. Besides an emotional cost, sometimes data can be used against us to cause a monetary loss. We might have a genetic condition that would prevent us from getting life insurance; we might get higher quotes from our plumber if he knew the balance of our bank account; or we might receive offers for risky financial in-

vestments right after we left a bar where we consumed one too many drinks. In a business-to-business setting, we might worry that our data will leak to competitors, or that our supplier, after finding out that we have a severe shortage, will use this information against us by raising prices.

3. Personal information might also be exploited by criminals. When we book a flight to Hawaii and this information becomes public, we basically put up an Open for Business sign for the local burglar community. Likewise, if our Social Security number falls into the wrong hands, we open ourselves up to the risk of identity theft.

Social stigma, various forms of discrimination, and outright criminal activity are all good reasons for protecting the privacy of those who have entrusted us with their data. This is true for all types of data, but it is especially important for data obtained as part of a connected strategy. One reason for this is that data created in a connected relationship tends to be richer, more current, and more confidential than data obtained in episodic relationships (if any data is collected at all in those interactions). Another reason is even scarier. Because of the automated nature of some elements of the connected relationship, customers are not only at risk of having strangers access their data, but hackers could also control the temperature of their houses, open and close the doors, control their connected cars, or steal money out of their bank accounts.

To create a connected strategy, trust between the firm and the customer becomes an essential element. A loss of trust will damage or end long-term relationships with customers very quickly. Data collection can either engender trust or destroy it. A core question you need to ask is, How do our data collection and usage affect our customers' trust in our company? Best to ask it early—and often.

When firms don't get the trust equation right, connected strategies can backfire. Remember the well-publicized story of Target inferring the pregnancy of a young woman from her buying habits and sending her coupons for maternity clothing, making her dad confused and angry because he didn't know about the pregnancy, or the question-able and unauthorized use of Facebook data by Cambridge Analytica to affect voting in the 2016 US presidential election, or various Google apps tracking location information despite a user's having turned off location history. Such missteps can be very costly and cause your customers to lose their trust in your ability to keep their data confiden-tial and to use it responsibly.

Moreover, firms developing novel connected strategies can find themselves in regulatory gray zones. Should drivers for ride-hailing companies be considered employees or independent contractors? Should staying in an apartment rented via Airbnb be considered an illegal short-term rental? Should recordings gathered automatically by Amazon's Echo devices be able to be requested as evidence in court cases? These are all legal issues that are currently being worked out. As you develop a connected strategy for your firm, you must stay abreast of the regulatory changes that affect you.

The Disruptive Potential of Connected Strategies

We covered a lot of ground in this chapter. In the next chapter, we will guide you through a workshop that will allow you to start apply-ing the first concepts of connected strategy to your own organization.

To summarize, we first introduced the concepts of willingness-to-pay and the efficiency frontier of an industry. The attributes of your product or service and the way you interact with your customer affect your cus-tomer's happiness, which translates into willingness-to-pay. At the same time, you incur fulfillment costs in trying to create this customer experi-ence. In other words, there is a trade-off between willingness-to-pay and fulfillment costs. Different firms will strike a different balance between willingness-to-pay and cost; they will position themselves dif-ferently in the marketplace. By plotting firms and their various posi-tions, we can identify the efficiency frontier in an industry. The efficiency frontier is defined by those firms who are furthest out in the willingness-to-pay/cost space. These firms are able to achieve the maximum willingness-to-pay given their level of fulfillment costs (or, conversely, they are able to minimize cost given their level of willingness-to-pay).

Connected strategies are so disruptive because they allow innovative firms to push out the efficiency frontier. From the perspective of other firms, it looks as if the firm with a connected strategy has completely broken the traditional trade-off between willingness-to-pay and cost. Firms with connected strategies are able to break the existing trade-off by fundamentally changing how they connect with their customers and whom they connect. Blue Apron reshaped how a customer thinks about grocery shopping for cooking: rather than finding a recipe, making a shopping list, and driving to various stores, a customer receives recipes and the exact ingredients at home. At the same time, Blue Apron created new connections between local farmers and end customers through its meal boxes. Likewise, ride-hailing companies redefined how customers interact with ride service providers. Rather than hailing a cab or calling a dispatcher, customers order a ride via an app and pay for it seamlessly. And ride-hailing companies created many new connections between previously unconnected parties: individual drivers with cars and riders with mobility needs.

While pushing out the efficiency frontier is a key step toward achieving competitive advantage, it is not sufficient. The main question is whether other firms can easily replicate your strategy and follow your move to the new position outside the old efficiency frontier. We will have more to say about this topic at the end of chapter 5, when we talk about the repeat dimension of connected strategies. As we will see, the repeat dimension holds a key source of sustainable competitive advantage.

Lastly, we touched on the crucial issue of data privacy, a topic we will repeatedly return to throughout this book. Connected strategies fundamentally rely on a trust relationship between customer and firm. Customers send, actively or automatically, data to firms with the expectation that firms utilize this data to create a superior customer experience. Failing to uphold this trust puts both the strategy and the firm at risk.

3

Workshop 1

Using Connectivity to Provide Superior Customer Experiences at Lower Costs

---

As promised, at the end of each part of this book, we have a workshop chapter. In this workshop, you will start exploring initial ideas for your connected strategy. In the workshop in chapter 6, you will consider what it takes to create a connected customer relationship. In the workshop in chapter 10, we will guide you through the process of creating a connected delivery model. At the end of chapter 10, we will also provide a summary of how the three worksheets form the basis for creating your overall connected strategy.

Please visit our website at connected-strategy.com for additional material, including larger templates for the exercises in all the workshops. These are easier to fill out than the pages in this book. We have also posted examples of filled-out templates that can be very useful as models when you start working through the exercises for your own organization.

At this point, we have not yet presented to you many of the tools to create a connected strategy. As a result, this workshop is shorter than the ones in chapters 6 and 10. Nonetheless, we want you to start thinking of what connected strategies might look like for your organization. We will help you do this by having you work through the following steps:

1. Ask a set of diagnostic questions concerning your current connections with customers.

2. Brainstorm the effects that a connected strategy could have for your organization.

3. Start identifying drivers of willingness-to-pay.

4. Sketch the efficiency frontier for your industry that reflects the trade-off between willingness-to-pay and fulfillment costs.

Step 1: Diagnostic Questions

We have found the following questions to be a good starting point for thinking about a connected strategy. These questions might appear simple, almost naïve, at first. As we found out in our discussions with many companies, however, they are surprisingly valuable.

- How often do you currently connect to your customers?

- What kind of information do you receive about your customers' needs?

- How does information flow from the customer to you? For instance, does the information flow rely on the customer taking the initiative, or does the information flow happen in a more continuous and autonomous manner?

- How long does it take for a customer need to reach you? (How much time elapses between a customer's realization that he or she wants or needs your product or service and the receipt of this information by your organization?)

- How long does it take for you to react once you have received a customer need?

- What do you learn each time a customer connects to your firm? How are you integrating these episodic interactions into a single, connected experience for your customers?

Step 2: Brainstorm the Potential of a Connected Strategy

Now let's get a bit more playful and start imagining what a connected strategy could do for your organization. The cases discussed in chapter 2 and the podcasts featured on connected-strategy.com might provide you and your team with some sparks to ignite creative discussions of the potential of connected strategies for you. The following prompts will also help you move from knowledge to action.

Imagine a world in which customers could instantaneously communicate their needs to you. You are by their side as they go through life, anytime and anywhere. How would this increase in connectivity allow you to improve how you serve your customers? More specifically,

- How could you use this information to increase the willingness-to-pay of your customers?

- How could you use this information to decrease your fulfillment costs?

Next, imagine a world in which you know a customer need even *before* the customer knows this need herself. Not only are you walking alongside her, but—given permission by your customer—you have visibility into her bank accounts and even some of the inner functions of her body. For example, your customer might need to save for retirement but has not given this matter any thought and presently is in debt. Or your patient might have a mild narrowing of the coronary arteries but is not currently showing any symptoms.

- How could you use this information to increase the willingness-to-pay of your customers?

- How could you use this information to decrease your fulfillment costs?

Step 3: Start Identifying Drivers of Willingness-to-Pay

The willingness-to-pay for a product or service is a result of three factors: how much a customer likes your product or service once he has it; how easy it is for him to obtain it; and how expensive it is to own the product. We will refer to the first part as *consumption utility*, the second part as *accessibility*, and the third part as the *cost of ownership*.

Consider consumption utility first. Consumption utility comes from various attributes of a product or service. For example, legroom (for air travel), the distance between two charges (for electric cars), weight (for bicycles), pixel count (for cameras), size (for clothing), and entertainment (for movies). The many possible attributes can be categorized into two groups:

Performance attributes: Performance attributes are features of the product or service that most people agree are better. For example, we all prefer more legroom, longer distances before recharging, lighter bicycles, higher pixel resolution, and so on.

Fit attributes: With some attributes, customers do not all agree on what is best. Some viewers like to watch *House of Cards*, while others might not. Some people enjoy a great steak, but vegetarians do not. When it comes to clothing, we have different body shapes and need differently sized clothes. Hence the term *fit* for these attributes.

Next, consider the accessibility of the product or service. Customers often face an inconvenience in obtaining the product or receiving the service. Economists often refer to this component as *transaction costs* or *friction*. Everything else being equal, we prefer our food here (as opposed to three miles away) and now (as opposed to after a thirty-minute wait). The following are the two major components of accessibility:

Location: How far does your customer have to travel to get your product or service?

Timing: How long does your customer have to wait for your product or service?

The third and final component of willingness-to-pay is the cost of ownership. As customers, we derive more value from a product that lasts longer and thus is cheaper on a per-usage basis. No matter whether it's electronics, sports gear, furniture, or kitchen equipment, products wear out, become obsolete, or need to be replaced for other reasons. Other elements of the cost of ownership include the need for maintenance and repair, benefits from potential warranties, and everything else that customers need to pay for when using the product.

Worksheet 3-1 summarizes these dimensions of willingness-to-pay and will help you keep track of the relevant drivers in your business environment. The drivers of willingness-to-pay are the most important variables from the perspective of your customer. Some of them are obvious. We all want greater products or services right here, right now, and, ideally, free of charge.

Don't stop there. At this point, consider the work that you did as part of step 2 when you brainstormed about immediate information availability and ways to know a customer's needs before even the customer becomes aware of the need. By having a connected relationship

Identify willingness-to-pay drivers of your customers

Consumption utility: How happy is the customer with the product or service?

Accessibility: How easy is it for the customer to get the product or service?

Cost of ownership: How much does it cost the customer to use and maintain the product?

Performance	Fit	Location	Timing	Usage costs over product life	Maintenance costs over product life

with your customer, including instantaneous communication, you can hopefully identify more subtle needs (and thus alternative drivers of willingness-to-pay) than "right here and right now." For example, you might realize that your customer is afraid of planning for retirement because this would require difficult discussions with his spouse, or that your patient really felt sick multiple times before showing up to the emergency room but did not bring this up with the primary care physician during a regular visit because there wasn't enough time. Those are the nuggets that will be the starting point for designing a connected customer relationship, which we will help you do in the second workshop.

Step 4: Sketch the Efficiency Frontier for Your Industry

Companies cannot be good at everything. They face trade-offs in their business. For example, they trade off performance attributes and the costs of providing products or services. More legroom means higher willingness-to-pay but fewer passengers per plane and higher costs per passenger. Similarly, there is a trade-off between the cost of providing goods and services and how inconvenient it is for a customer to access them. As passengers, we would love to have a car waiting for us at every corner of town—it would translate into convenient access. But it also would mean lower vehicle utilization and higher costs.

Such trade-offs can be illustrated graphically using the efficiency frontier framework discussed in the previous chapter. Worksheet 3-2 helps you draw the efficiency frontier in your industry through the following steps:

1. Pick your most relevant competitors.

2. Rank them in order of the willingness-to-pay that their products or services create. In other words, put yourself in the shoes of a typical customer and ask yourself which product or service is the most desirable, ignoring the price you would

WORKSHEET 3-2

The efficiency frontier for your industry

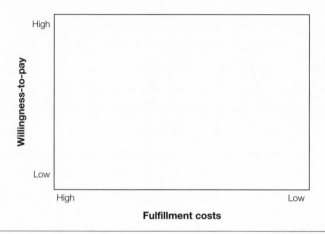

have to pay for it. For example, if someone else paid for it (though you would still have to do the purchasing, transaction processing, and waiting), would you rather receive a Mercedes C-class, a BMW 328, a Tesla Model 3, or a Lexus IS? You can refer back to step 3 of this workshop and ask yourself how well your competitors perform along the dimensions of performance and fit attributes, location, timeliness, and cost of ownership. An accurate, absolute measurement is not required—relative comparisons are sufficient. If you believe that the willingness-to-pay is very different for different market segments, draw a separate efficiency frontier for each market segment.

3. Rank your competitors and yourself in terms of fulfillment costs. Ask yourself, What is the average cost per transaction that you have versus your competitors? Note that this is the average cost. For instance, one firm might spend a lot on advertising, but if it has many customers, it can spread this cost over many transactions, affecting its average cost per transaction only by a little. Again, a simple rank ordering is sufficient at this time.

THE COMPETITION BETWEEN POWER TOOLS AND NECKTIES

It's easy to ask, "What are your most relevant competitors?" But that seemingly simple question is much more complicated than one might initially suspect. Consider the following example, inspired by a remark on substitutes by Michael E. Porter, one of the most influential scholars of modern strategy. Imagine you work for a power-tool company. As a power-tool company, who are you competing against? One way of answering this question is by looking at who else is producing power tools. Framed this way, competitors such as DeWalt, Bosch, Ryobi, Hilti, and similar tool makers might make it onto your list.

Another way of tackling this question is to reframe it and ask, "For what purpose do customers use power tools such as a circular saw?" Chances are your customers might want to use the saw to cut some wood, maybe in order to build a new dining table. When asked this way, the question reveals that you, as the maker of circular saws, are not just competing with other power-tool makers, you also are in competition with the makers of dining tables. The more customers are willing to tackle the job of building dining tables themselves, the more money they will pay to power-tool companies

4. With this information, you can position yourself and your competitors in worksheet 3-2. Next, draw the line that represents the efficiency frontier. This is the group of firms that are farthest up and to the right on the plot—that is, the firms that have the highest willingness-to-pay for any given level of fulfillment costs (or, conversely, that have the lowest cost for any given level of willingness-to-pay).

Once you have filled out worksheet 3-2, you can think about the following questions:

versus furniture stores. (Companies making connected power tools have taken advantage of this by using connectivity to make it easier to use their tools. Check out our website for a podcast featuring an interesting discussion on connected power tools with industry leaders Hilti and Bosch.)

But we can reframe the question again and ask, "Why do people buy power tools in the first place?" When you ask the question this way, you will quickly realize that one of the most common reasons for power-tool purchases is that customers are shopping for Father's Day or Christmas presents. You probably sense where we are going with this argument. In this framing of the question, you realize that power-tool makers are competing with other providers of Father's Day or Christmas gifts, neckties being one of the most important ones.

So, when we ask you to think of your competitors, how broadly should you look? We propose that you start with those companies that you are directly competing against—in our example, this would be the other makers of power tools. But, especially when it comes to imagining innovation and disruption, we want you to think about the competition more broadly by considering the ultimate needs that your customers want to fulfill, a topic we will return to in chapter 5. Disruption typically happens from players outside your current market. Just remember: hotels were disrupted not by other hotels but by Airbnb . . .

- Where are you relative to the efficiency frontier? Are you on the efficiency frontier, or are there firms providing a similar (or even higher) willingness-to-pay while enjoying lower fulfillment costs? Recall our comparisons between meal-kit deliveries and farmers' markets and between ride hailing and cabs. Not every firm in an industry is likely to be on the frontier.

- If you are not on the efficiency frontier, what efficiency improvements do you plan to pursue in order to reduce your fulfillment costs?

- Assuming you are on the efficiency frontier, do you feel that you are in the right spot on the frontier? Or do you feel that you should rethink the trade-off between willingness-to-pay and costs (e.g., sacrifice some efficiency to provide a better product or service)?

- What are the trends in your industry? Is there a pressure to lower costs (moving to the right), or do you see your firm win over its rivals by providing products and services with a higher willingness-to-pay (moving up)?

- Are there new technologies that have allowed some of the firms already in the industry or potentially new entrants to push out the frontier? Do you see new business models breaking the trade-off between willingness-to-pay and fulfillment costs?

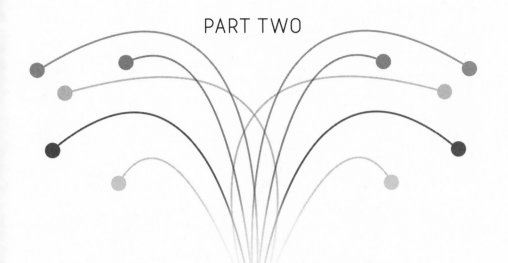

PART TWO

CREATING CONNECTED CUSTOMER RELATIONSHIPS

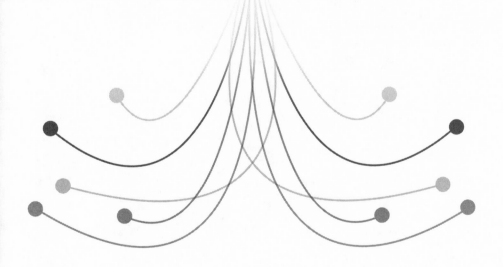

4

Recognize, Request, and Respond

Building Connected Customer Experiences

Here is a typical customer experience that you've probably come across: David sits in his home office, about to print a set of urgent letters, when his printer runs out of toner! He grumbles, gets into his car, and drives to the nearest office supply store. He wanders the huge store until he finds the toner aisle. Hopefully, he remembers the exact printer model he has. Is it the HP Office JetPro 6978 or the HP Office JetPro 8710? Each uses a different cartridge, even though they look pretty similar. He decides it is the 6978 (turns out it's his lucky day and he is right). He grabs a multipack because the store is out of single cartridges, and he prays that his aging printer will last long enough to use all this toner. To pay, he waits in the long checkout line. When it's finally his turn, he has trouble finding his credit card. He eventually pays, gets back into his car, and makes it home nearly two hours later to resume printing.

Now, let's switch perspectives. Assume you are in charge of the toner division at a printer company. What are you spending your resources on? Most likely, a good proportion goes to developing longer-lasting toner, crisper colors, better cartridges, and efficiency improvements from manufacturing to distribution. While there is nothing wrong with a firm's striving to build better products or to drive down costs, our printer example shows how far removed those actions often are from the real pain points of the customer. Customer and firm are really poorly connected.

In this and the next chapter, we dive into the concept of a *connected customer relationship*. These two chapters are followed by a workshop chapter, which will help you build your own set of connected relationships. As we defined in chapter 1, a connected customer relationship is a relationship between a customer and a firm in which episodic interactions are replaced by frequent, low-friction, and customized interactions enabled by rich data exchange. What would this look like in practical terms? It surely must be different from David's experience with the toner.

Let us first look at one experience at a time. We want to understand the process that a customer goes through of *recognizing* a need, *requesting* a solution to this need, and experiencing how the firm *responds* to the request. The next chapter adds the fourth R of connected relationships, *repeat*, and will explain how such individual customer experiences are woven together over time to create a lasting customer *relationship*.

Like any customer experience, the interaction in our toner example starts with a customer need. The printer uses up toner one page at a time. As long as the printouts are of high quality, the customer might not be aware of the need for more toner, even though the printer cartridge might be almost empty. But as the ink on the paper becomes weaker and weaker, the customer gains awareness of the need. In marketing language, needs of which customers themselves are not yet aware are called latent needs.

Once customers are aware of their needs, they search for options. This step can be remarkably complex, especially these days, with our ability to order almost anything from anywhere around the globe. Quite often customers will not be aware of all the options that could possibly satisfy their needs. Once options have been surfaced, a customer will have to choose from among retail channels, brand names, product quantities, and much more. Given the complexity of options, deciding which is best isn't easy. Finding the product, paying for it, and taking it home is a time-consuming hassle.

But the interaction episode is still not complete. Our customer still has to replace the old toner cartridge with the new one. Lots of things can still go wrong at this stage—the new cartridge might leak or the plastic hinge in the printer might break, and rather than printing his letters, David is now on the phone talking to technical support.

As the printer example illustrates, even one single interaction episode between customer and firm can be a long journey. While we described this journey for an individual end customer, it is important to note that customers in business-to-business settings experience a very similar journey. We find it helpful to break up this customer journey into three distinct phases, each dealing with a fundamental question:

1. Why does the customer engage in the interaction in the first place?

2. How does the customer go about identifying, ordering, and paying for the desired product?

3. What products and services are provided to the customer?

Each of these three phases has a number of steps to it, as is illustrated in figure 4-1.

Moreover, each of these three phases corresponds to a key design dimension of connected relationships. As a manager building connected relationships, you need to ask yourself how you will *recognize* your customers' needs (or help them recognize their needs); you need to configure activities that will help the customer identify and *request* the option that would best satisfy this need; and you need to put in

FIGURE 4-1

The three phases of the customer journey

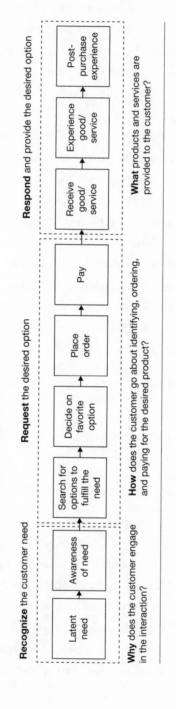

place a system that allows you to *respond* with this desired option in a cost-effective way.

We will use the printer example to illustrate the differences among the four connected customer experiences that we presented in chapter 1:

Respond-to-desire

Curated offering

Coach behavior

Automatic execution

As we shall see, these four experiences start affecting the customer earlier and earlier in the customer journey. Let's go back to David's experience buying a printer cartridge. To state the obvious, David's experience was not particularly positive and reflected the total absence of any connection between him and the printer company. Though not a formal term, we might label such an experience a buy-what-we-have customer experience. The firm waits for the customer to show up, and the customer has to buy what's available. Practically the entire customer journey was David's responsibility. David had to realize the need for a new toner cartridge. David had to figure out which precise toner he needed. David had to drive to a store, search for the product, stand in line, pay for the product, drive home, and replace the cartridge. How can we improve this customer's experience?

The Respond-to-Desire Connected Customer Experience

How do you think David would feel about the following customer experience? Upon realizing that he needs a new cartridge, David goes online to his favorite retailer, types in his printer model, clicks to order the correct toner, and pays with the same click because his credit card number and shipping address are already stored. Two hours later, his doorbell rings and the toner is delivered. Though David is still in the driver's seat for executing the transaction and performing almost every step of the customer journey, this transaction is associated with far less friction, so it is much more pleasant for David.

This is the idea behind a *respond-to-desire* connected customer experience: firms try to provide the customer with the desired product or service in the fastest, most convenient way possible. In terms of the customer journey, firms using respond-to-desire significantly reduce the friction from the moment a customer has decided on the desired option to the moment the customer receives the product. Moreover, these firms are able to provide the customer with exactly the desired option (only a single black toner cartridge, not the multipack). To understand the anatomy of a respond-to-desire connected customer experience, consider figure 4-2.

The figure divides the steps along the customer's journey from latent need to the postpurchase experience that we introduced in figure 4-1 into two spheres. The upper half of the figure shows steps that are carried out by the customer, while the lower half captures steps that are carried out by the firm. Our toner example started with the customer realizing the printer had run out of toner. Note that the following steps have to happen regardless of whether the customer buys replacement toner online or goes to the local office supply store: the customer had to look at some available options (and avoid the confusion between different model numbers), the customer had to choose, and the customer had to place an order, be it at the physical or the virtual checkout.

From the perspective of the customer, what makes for a good respond-to-desire experience? One element is clearly the amount of effort expended; the less, the better! One-click checkout with a credit card on record is more convenient than going to the store. Customers like companies that listen carefully to what they want and respond to their desires quickly. Hence the name respond-to-desire.

For some respond-to-desire customer experiences, such as running out of toner, speed is the most important attribute the firm must mas-

FIGURE 4-2

The respond-to-desire connected customer experience

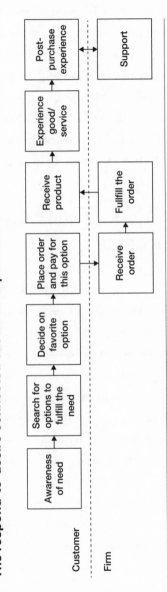

ter. For instance, being able to send a car within minutes is a critical attribute for customers of Uber or Lyft. For Airbnb, in contrast, the critical attribute affecting a customer's happiness is the variety of options in style and price where the customer wants to visit. The speed of booking is important but not the key advantage propelling the use of the service.

Amazon has been a pioneer in combining the order and payment process into "one click." To further facilitate ordering, Amazon introduced Dash buttons, small Wi-Fi devices that can attach to the fridge, the washing machine, or the bathroom vanity. With the press of a button, items from bottled water to baby wipes can be reordered.

The Curated Offering Connected Customer Experience

How could the online toner-shopping experience be improved? Imagine that, after David logs in to his online account, the site suggests the correct toner cartridge based on his prior purchases, which eliminates the need for him to figure out the right type of toner for his printer. In addition, the site could also suggest reordering paper, even though David did not ask for any paper (but is about to run out—good catch!). This is what we call a *curated offering* connected customer experience. Amazon, for example, shows a "frequently bought together" selection and is able to make personalized recommendations, taking into account what the customer has purchased in the past and what bundles of items are often purchased together in the same transaction.

Respond-to-desire requires the customer to articulate a specific need. Sometimes, however, customers might not know exactly what they want, or it would take them considerable effort to figure out. In the curated offering connected experience, the firm becomes active in the customer journey earlier than in the respond-to-desire experience. While respond-to-desire requires customers to know exactly what they want, curated offering helps customers at earlier stages in the journey, searching for options and deciding on options. Take another example: a customer knows she wants to watch a comedy but might not know what new comedies have been released. In this case, a curated offering is very helpful. Netflix suggests comedies based on previous choices and choices of similar customers. In general, curated offering makes search outcomes much more personalized.

A curated offering thus customizes the firm's response with a specific set of options. In doing so, this format can also anticipate needs. Figure 4-3 shows the sequence of events for the curated offering experience. Again, the activities performed by the customer are on top, with the firm's activities on the bottom. As we can see by comparing figures 4-2 and 4-3, in the curated offering model, the firm takes a more active role in the search and selection process. Rather than expecting to receive a specific order, the firm surfaces options the customer wasn't aware of and makes recommendations concerning which options might best fulfill the customer's needs.

Blue Apron, or one of the similar meal-kit providers we discussed in chapter 2, is a powerful example of curated offering. Think about how Blue Apron's curated offering differs from a respond-to-desire experience. A respond-to-desire experience would have saved you from needing to run an errand by enabling you to order online and have products delivered to your house. In fact, home delivery services for groceries have witnessed dramatic growth. All of them share the principle of "you order, we deliver." Though this is better than spending your time in a supermarket checkout line, the burden of finding recipes, choosing one, and creating the shopping list still lies with you as the customer. And you are still stuck with the extra ingredients that did not fit the recipe's portion sizes. In contrast, for many customers, relying on Blue Apron to organize the meals and portion sizes is more convenient, more fun, and also healthier, creating a higher-quality service compared with a respond-to-desire experience, not to mention a buy-what-we-have approach. (For another example of curated offering, see the sidebar.)

FIGURE 4-3

The curated offering connected customer experience

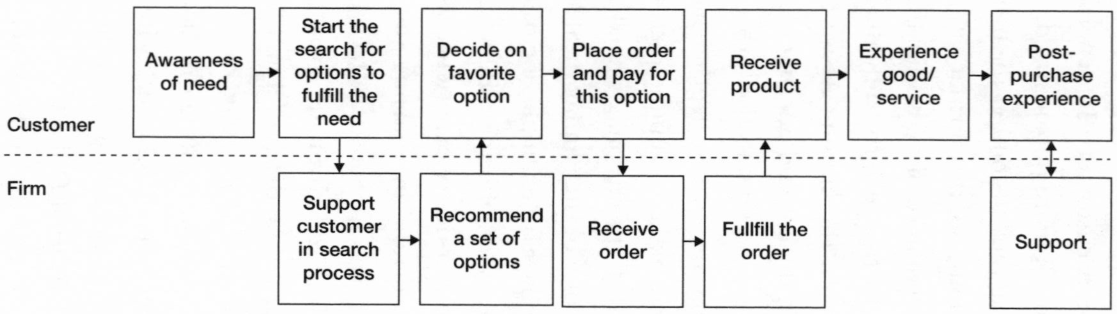

TRUE PERSONALIZATION

Curated offering experiences have been helped tremendously by advances in production and sensing technologies that allow customization down to the individual. The cosmetics industry was one of the early adopters of this development. The market for foundation, which is around $1 billion in the United States alone, is a good example. With all the variations in skin tone and texture, the quest for the right shade and formulation of this cosmetic has been a major pain point for customers for a long time. With Shiseido's bareMinerals Made-2-Fit foundation, a customer can use her smartphone to download an app, scan her face, and receive a personalized foundation that matches her skin color.

Advances in 3-D printing technology also promise unprecedented customization with respect to pharmaceuticals. With pills, for example, inventory constraints limit the number of dosages that are currently available. With 3-D printers, tablets or gummies can be created directly at the pharmacy with the desired personalized dosage. The created pills can even contain more than one needed medication (so-called poly pills), or gummies can be made in more interesting-looking shapes (dinosaurs, trucks, etc.) for children.

The Coach Behavior Connected Customer Experience

Returning to our printer example, neither respond-to-desire nor curated offering solved a basic problem: the customer realized the need to buy new toner only after the toner ran out. The firm waited for the customer to take the initiative, to start the interaction. Unfortunately, in many cases, customers are late in taking the initiative, causing inconvenience.

How can we help customers remember earlier? Perhaps the retailer could have already sent the customer a reminder to reorder last week. Such a reminder might have been based on past purchasing behavior—the customer bought toner roughly every eight weeks, and now it has been seven weeks since the last order. While reminding the customer to reorder toner, perhaps the retailer could also remind the customer to run the cleaning function on the printer to keep print quality high—another action customers know they should be doing but don't. Either way, the firm serving the customer is more proactive than in the previous two connected customer experiences.

We call this type of connected experience *coach behavior*. There are many instances in which people would like to engage in certain actions, but inertia and decision biases get in the way. We would like to lose weight, but we have a hard time sticking to a healthier diet. We want to become more fit, but we can't stick to our workout regimen. We need to take our medications, but we are forgetful. With coach behavior, the firm takes on a parental role and coaches the customer to change behavior. In the customer journey, it acts one stage before curated offering kicks in. It activates customers' awareness of their own impending needs.

In figure 4-4, we have illustrated the coach behavior connected customer experience. Since most of the time it is a behavioral change that is desired (rather than the purchase of a discrete good or service), we have changed the labels in some of the steps. Rather than recommending a set of options, the firm recommends that the customer take a particular action (or reminds the customer to do so). The customer then decides which action to take, and ideally behavioral change ensues.

Notice how the customer experience in figure 4-4 puts more emphasis on the firm taking action. While the customer ultimately takes action (takes the pill, avoids the burger, goes to the gym), the firm watches over the customer, knowing what the customer needs now and in the long run—not just what the customer wants right now. As with

FIGURE 4-4

The coach behavior connected customer experience

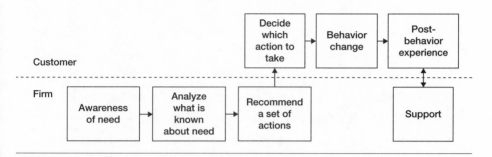

the curated offering, such knowledge might come from observing many customers, or from what the customer has previously told the firm. Imagine a coach behavior experience in which a customer, to maintain a diet, tells his bank to block his credit card whenever he is within one hundred feet of a fast-food restaurant. In a respond-to-desire customer experience, the customer would get that cheese-burger whenever he desires, in the most convenient way. But the customer also knows that he would regret it twenty minutes later. So, a good coach behavior experience is indeed somewhat parental.

We know that creating behavioral change can be difficult; other-wise, all of us would easily stick to our New Year's resolutions. Peer pressure is a potent tool in this regard. As a result, a number of firms offering coach behavior experiences also facilitate the creation of peer-to-peer networks in which participants can celebrate their efforts ("I did 12,000 steps today!") and encourage each other. By borrowing tools developed by game designers to make online games engaging, if not addictive, many firms are using "gamification" to create behav-ioral change. Participants collect points and badges, and engage in friendly competition against themselves and peers to accomplish their behavioral goals. We will come back to these networks in chapter 7. (For more examples of coach behavior, see the sidebar.)

COACH BEHAVIOR WEARABLE SENSORS

Wearable sensors are enabling a whole range of coach behavior customer experiences. Consider the following three examples. L'Oréal has developed a battery-free wearable sensor to measure individual UV exposure. The sensor is less than two millimeters thick and nine millimeters in diameter, and it is designed to be worn on the thumbnail. It can store up to three months of data and is paired to a smartphone app to collect and display daily UVA and UVB exposure, offer personalized sun safety coaching tips, and track trends in UV behavior over time.

For individuals who want to learn or improve their yoga poses but don't always have access to an instructor, Sydney-based Wearable X has created a solution. Its Nadi X pants have woven-in sensors around the hips, knees, and ankles that measure body positions and provide haptic feedback, guiding the athlete to the correct positions through gentle vibrations. When paired via Bluetooth to an app on the customer's phone, further visual and audio cues break down different yoga poses that supplement the vibrations.

To help runners avoid injuries by coaching them to achieve a better running style, Sensoria developed an anklet device attached to a special running sock that users wear while running. Sensors in the bottom of the sock measure where the user's foot makes contact with the ground and for how long. The anklet contains a CPU that analyzes a rich variety of data from the sensors and relays that data to an accompanying smartphone app. The app displays a detailed heat map of where pressure is being placed on the foot, along with detailed statistics regarding foot contact time, cadence, steps taken, stride length, and speed. This information is shown in the Sensoria app so users can adjust their movements in real time. The app also gives audio alerts in real time with an automated voice that provides feedback about any incorrect foot-strike patterns or other adjustments to prevent injury.

The Automatic Execution Connected Customer Experience

Let's revisit the printer story one more time. This time, imagine that the printing is going well when the doorbell rings. David is surprised to see a box delivered. He doesn't recall having ordered anything. Inside the box is a toner cartridge. How odd, he thinks. He resumes printing, and his computer alerts him that his printer is about to run out of toner. Only then does he remember that when he bought the printer, he gave the printer company permission to automatically send more toner as it ran low. David just experienced an *automatic execution* connected customer experience. Once the firm is authorized to take care of something, the firm automatically gathers information and fulfills the need, often before the customer had realized that the need has arisen. This is shown in figure 4-5. This diagram is almost the opposite of the respond-to-desire flow in figure 4-2. Here, almost all activities are controlled by the firm because the firm knows what the customer needs and when. This is somewhere between Big Brother and loving mother.

Delivering such a connected customer experience can be difficult. Unlike in the case of curated offering, the lack of customer involvement in the decision-making steps of the customer journey makes errors more possible. True, customers who go online and search for a book on baby names are likely to be future parents. But should a retailer send them a crib and diapers based on this information alone? Making automatic execution connected customer experiences work requires increasing the bandwidth of the information flow that moves from customers to firms.

With the increasing connectivity of objects through the Internet of Things, more and more of these automatic relationships based on continuous information flows will become possible. The printer example is real. HP has a program called Instant Ink that works exactly as we

FIGURE 4-5

The automatic execution connected customer experience

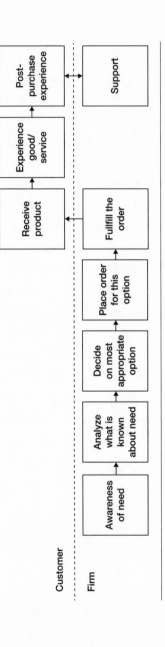

described. In this case, HP ships replacement toner to customers once their printers send out a "low ink" signal to the company. Brother has a similar program called Brother Refresh, where customers can decide whether they want the company to send the ink or leave the fulfillment to Amazon.

Soon our refrigerators, sensing that the weight of the milk container is very low, will be able to reorder our milk for delivery by tomorrow morning. (Of course, the fridge also checked our calendar to make sure we were not going on vacation starting tomorrow and wouldn't need the milk.)

Other already-existing examples of automatic execution connected customer experiences are in the realm of medical intervention. Fall-detection sensors are small medical devices for seniors who are at increased risk of falling. Early generations of this technology followed the principle of respond-to-desire. A senior in need of help could press a button on a wearable device, activating an emergency call anywhere in the apartment or house. This was clearly addressing an important unmet need, as previously, a senior who fell down the basement stairs might have needed urgent medical attention but was unable to move and reach the phone. That is why the latest fall-detection devices have switched to an automatic execution relationship. Sensors in the device detect the fall and are able to take action automatically, even without the involvement of the patient. (For another example of merging curated offering with automatic execution, see the sidebar.)

The Information Flow from Customers to the Firm

A central aspect in creating new connected customer experiences is to design the information flow between the customer and your firm. After all, it is this information that allows you to recognize your customer's

AUTOMATIC EXECUTION WITH VIDEO GAMES

Consider the following stunning statistic: if the players of the hugely popular online game *World of Warcraft* spent their time creating Wikipedias rather than playing the game, they could create a new Wikipedia every week! That's a remarkable testament to how engaging and "sticky" games have become. Video game companies have come a long way when it comes to curated offering and automatic execution. It used to be that every player would purchase the same version of a game by going into a store and buying a game cartridge or CD. Now, with online gaming, game producers learn about the preferences and skills of individual gamers and create customized experiences that keep a player in a state of flow. Players are exposed to challenges that are neither too difficult (causing frustration) nor too easy (causing boredom), a process called dynamic difficulty adjustment that requires sophisticated artificial intelligence on the back end. Likewise, different players derive pleasure from different aspects of a game. There are achievers (interested in gaining levels and points), explorers (interested in understanding the nuances of the game), socializers (interested in interacting with other players), and killers (the name says it all). By understanding a particular player's type, the video game can adjust and get the player into more enjoyable situations. No wonder one study of *World of Warcraft* found that 75 percent of gamers play longer than two hours per day on average, and 25 percent play longer than five hours per day.

needs and to identify the optimal solution. We find it helpful to think about five dimensions to describe this information flow:

1. The trigger of the information flow, which could be the customer or the firm

2. The frequency of the information flow (episodic vs. continuous)

3. The richness or bandwidth of the information flow (low vs. high)

4. The customer effort associated with this information flow (low vs. high)

5. The information processing that is required to infer the right product or service solution in response to the customer need (the customer might explicitly express which product or service she wants, or the appropriate product might be inferred by the firm)

You need to make decisions along these five dimensions before you can turn to the technical aspects of building a connected strategy, such as smarter devices or increased communication bandwidth. This is summarized by table 4-1.

TABLE 4-1

Information flow dimensions for different customer experiences

	Buy-what-we-have	Respond-to-desire	Curated offering	Coach behavior	Automatic execution
Trigger of information flow	Customer	Customer	Customer	Firm	Firm
Frequency of information flow	Episodic	Episodic	Increased	Continuous	Continuous
Richness of information flow	Low	Low	Medium	High	High
Customer effort to send information	High	Reduced	Reduced	None	None
Information processing to find preferences	All preferences expressed explicitly by the customer	All preferences expressed explicitly by the customer	Firm makes recommendations to the customer	Firm attempts to drive very specific customer action	Firm autonomously takes action for the customer

At the beginning of the chapter, we introduced the buy-what-we-have customer experience, illustrated by the customer's painful journey from the late realization of toner shortage, to a painful ordering process, and finally to the reception and usage of the product. This experience was not part of a connected relationship. The episode was triggered by the customer's realization that the printouts were of poor quality. The only information flow took the form of a customer purchase (and even that took a detour by going through a retailer). And it was left entirely to the customer to decide that purchasing toner for the JetPro 6978 was the right solution for his printing needs.

The respond-to-desire connected customer experience was primarily one of reducing the friction associated with the transaction. This could be accomplished via a simple ordering interface (e.g., one-click shopping), conveniently located sensors (e.g., Amazon Dash buttons), or voice recognition (e.g., Google Assistant). All of this dramatically reduces the customer effort compared with ordering via cumbersome websites, telephone, or mail or making a trip to the store.

While the curated offering connected experience also relies on the customer as the trigger of the transaction, it puts the firm in a more active position of helping the customer figure out a solution to her needs. Central to this is the recommendation process. Based on past purchases (the customer spends two hours per day streaming World War I documentaries) or expressed preferences ("I really like Bollywood movies"), Netflix can make recommendations to the customer on how to fulfill her entertainment needs. This recommendation process requires data about the customer, increasing the need for information frequency and information richness. The more the firm knows, the better a recommendation it can make. Technical challenges for this connected relationship include a strong recommendation technology. Moreover, firms need to deal with the demand for privacy by customers, as became obvious when Google was criticized for reading through its customers' Gmail accounts to improve targeted advertising.

Because overcoming customer inertia is one of the main benefits of the coach behavior connected relationship, we cannot rely on the customer to be the trigger for transactions. Instead, the trigger point is moved to the firm. That has consequences for the other dimensions as well. The firm needs to be receiving information from the customer all the time so that it doesn't miss the right moment to take action. In this case, the connection to the firm is always on, and customers send information to a firm continuously and quite often autonomously. You could say the firm is automatically "hovering" over the customer. For instance, a customer's Fitbit is continually gathering information, using the customer's smartphone as a relay device to automatically send information to a health care provider or the customer's personal trainer. The technical challenge for this connected relationship lies in enabling cheap and reliable two-way communication between customer and firm, reflecting the fact that communication in this case has to happen 24/7.

Finally, as we move to automatic execution, the firm takes on all responsibility for finding the right solution to the customer's needs. There is an important difference between a fall sensor that waits for its user to press an alarm button and a device that issues a distress call based on the reading of an accelerometer. The sensor has to be read continuously and ideally transmits information in real time so that even in the rare event that the device is damaged in a fall, help can still be sent without delay. Technically, this raises the bar for the ability to correctly infer the right product or service for the customer based on the automatically transmitted information. Recommending a bad movie or reminding a male patient to make a gynecology appointment might lead to customer annoyance. Shipping a Brother cartridge to the user of an HP printer, in contrast, has more significant costs associated with it, but even these costs are small when compared with the damage done by a failure to call a much-needed ambulance.

ARTIFICIAL INTELLIGENCE AND DEEP LEARNING

As we have seen in our discussion in this chapter, there often is more to con-nected customer relationships than just providing customers with what they ask for. Curated offering, coach behavior, and automatic execution all rely on joint problem solving between the customer and the firm. In the past, such joint problem-solving behavior was only possible through human experts, oftentimes sales managers who would help the customer to de-termine which product or service was right for them and when and how to acquire it.

Thanks to advances in artificial intelligence, human skills can now be vastly augmented and more and more processes can be automated. The field of artificial intelligence is concerned with equipping machines with skills that previously could only be mastered by humans, thereby enabling them to not just execute orders but solve problems instead.

It is helpful to distinguish between two ways a computer can help in the customer journey. The first one is based on applying a (potentially large) set of rules, such as "propose paper to everybody who has pur-chased toner" or "reorder milk whenever the last bottle has been opened." Such rules can be carefully audited, making sure they lead to desired recommendations, but they have an important drawback. Every aspect of the problem needs to be coded in the form of a rule, which can quickly lead to an explosion of rules (e.g., if customers leave for vacation, a query of the calendar might also be required, so that milk is not reor-dered in this case).

Coding everything in rules might work for low-complexity situations such as ordering toner or milk. In situations of high complexity, however, it is much harder to codify knowledge in the form of rules (what exactly makes one skin irregularity an indicator of a risk of cancer?). This is where the second way in which a computer can solve a problem comes in. Rather than defining rules for identifying skin cancer, the computer gets fed a very large number of skin images and is informed which are cancerous and which aren't. Based on patterns the computer finds in these old images (often referred to as the training set), the computer then evaluates future images.

This approach is much more humanlike—after all, we do not tell our children what exactly makes an animal a cat; they just figure it out from observing many animals and listening to our parental classification. (It turns out that determining whether a video includes a cat has been a much-studied problem in computer science and has attracted the attention of some of the greatest minds in that field.) The technical term for this learning approach is *training a neural network*.

Such digital representations of neural networks have been around for many decades. Recently, a breakthrough in this area has been achieved, referred to as deep learning. Deep learning is inspired by the human brain and organizes the neural network into multiple layers, each layer using different levels of abstraction. For example, when looking for a cat in a digital image, the first level might identify the pixels on the image that constitute the edges separating one object from another. The second level might be concerned with the task of translating the edges into objects (such as the leg of the cat, the ears of the cat, or the couch under the cat). The third level might group the objects, and the fourth level might determine whether a group of objects is a cat. This approach requires lots of data and computing power, but it does not need prespecified rules.

Because of the repeat dimension in our framework (see the next chapter), firms pursuing a connected strategy have access to the data required for deep learning, making them a better partner in joint problem solving with the customer than their (unconnected) competitors.

The Different Domains for Different Connected Customer Experiences

While we are excited about the emergence and possibilities of new customer experiences related to automatic execution, we want to stress that we do not see them as the best solution to all problems or for all customers. In other words, connected relationships are *not* always becoming better as you move from the left to the right in table 4-1.

Customers differ in the degree to which they feel comfortable if things begin to happen automatically around them. An experience that is magical for one person might be creepy to another. One person might find it delightful that Disney sends them an automatically created picture book of their last visit to Disneyworld. Another customer might find it invasive. Not only do customers differ in what value drivers (or pain points) are particularly salient to them, but they also differ in the degree to which they are comfortable with sharing data and having the environment around them act on that data. Understanding which connected customer experience is best suited for a particular customer is as important as understanding the particular needs of this customer. Transparency and the ability of customers to opt in or opt out is key in this respect. Unless you have asked the customer whether it is OK for you to collect data, and unless you have explained very clearly how you will use this data and how this use will create value for the customer, you run the big risk of alienating rather than delighting your customers.

In sum, each of the four connected customer experiences has its own merits and works well in specific-use cases and for particular customers:

> ***Respond-to-desire*** works best when customers know what they want and the firm is capable of providing it quickly. The problem is that fulfilling a random customer request, such as "I want to eat a bacon cheeseburger now, even though I am in a

vegetarian restaurant and it is three o'clock in the morning,"
can be costly or impossible. The firm's essential capability is
an operational one: fast delivery, flexibility, and exact execution. Customers who like to be in the driver's seat, having full
control, like respond-to-desire.

Curated offering makes sense when customers don't know exactly what they want because they don't know all the available
options. In this setting, a firm can delight its customers by finding them a product best suited to their needs, and also gain efficiency benefits by proactively steering them toward something
that can be easily provided. The key capability here is the recommendation process. Customers who like to make the final
decision but still value advice benefit from curated offering.

Coach behavior is of the most value for latent needs that customers are aware of but have a hard time pursuing themselves,
because of inertia or some other behavioral reason. Yes, the
customer wants a bacon cheeseburger, but once reminded of his
cholesterol levels, he is willing to order a salad. For this to work,
the firm needs to have a deep understanding of customers'
needs. This is often based on a rich information flow from the
customer to the firm via automated hovering. It also needs to
balance keeping the customer engaged and loyal with being
parental and restrictive. Customers who do not mind sharing personal data if they see a clear payback in terms of being able to
achieve personal goals are willing to engage in a coach behavior customer experience.

Automatic execution should be the connected relationship of
choice only if the firm is able to understand the user so well
that it is better positioned to make purchase (or other) decisions
than the user herself. It also requires a setting in which
mistakes are not too consequential. Customers who are comfortable with having a continuous data stream from themselves

TABLE 4-2

Domains of application for the four connected customer experiences

Type of connected customer experience	Description	Key capability	Works best when	Works best for
Respond-to-desire	Customer expresses what she wants and when	Fast and efficient response to the incoming orders	Customers are knowledgeable	Customers who don't want to share too much data and want to be in control
Curated offering	Firm offers a tailored menu of options to the customer; final choice rests with the customer	Making good recommendations	The set of options is large, potentially overwhelming for the customer	Customers who are fine sharing some data but still want to have final say in decisions
Coach behavior	Firm nudges customer to adjust her immediate preferences, either to obtain some larger goal or to reduce fulfillment costs	Understanding latent customer needs and how they are (not) related to the immediate actions	Customers have inertia and other biases keeping them from achieving what is best for them	Customers who do not mind sharing personal data and an environment that influences their behavior
Automatic execution	Firm monitors, then executes the fulfillment process without action from customer	Monitoring the customer and translating incoming data into action	Customer behavior is very predictable and costs of mistakes are small	Customers who do not mind sharing personal data and having firms making decisions for them

(or their devices) to a firm and who trust that the firm uses the data to fulfill their needs at a reasonable cost will be the most open to an automatic execution customer experience.

Table 4-2 summarizes the connected experiences alongside the domains for which they are best suited and the capabilities a firm needs in order to create these customer experiences.

Recognize-Request-Respond: A Broader View of Satisfying Customer Needs

When we ask managers to list the drivers of willingness-to-pay of their customers, their main focus is usually on tangible and intangible aspects of their products or services, such as quality attributes and brand. Obviously, these are important factors, but the willingness-to-pay of a customer can be influenced by a much broader set of drivers. Every transaction your customer has with you is actually an entire journey, and at every step of this journey there is an opportunity to either delight your customer or have your customer suffer a pain point. We find it helpful to distinguish three phases of the customer journey: recognize—the part of the journey where a latent need of the customer arises and either the customer or the firm is made aware of it; request—the part of the journey where the need is translated into a request for a solution to the particular need; and finally respond—the part of the journey where the customer receives and experiences the solution.

Our research into connected strategies has revealed four distinct approaches that firms use to reduce the friction of this customer journey—in other words, four different connected customer experiences. These customer experiences are distinguished by what part of the customer journey they affect. The respond-to-desire connected customer experience starts at the point in the journey when a customer knows precisely what he or she wants. The firm's goal is to make it as easy as possible for the customer to order, pay for, and receive the desired product in the desired quantity. The curated offering customer experience acts further upstream in the journey by helping the customer find the best possible option that would fulfill his or her needs. Both respond-to-desire and curated offering can only work if customers are aware of their needs. Firms creating a coach behavior customer experience help their customers at exactly that part of their journey: they raise awareness of needs and nudge the customer into action. Lastly, when the firm is able to be aware of a customer need even before the customer is aware of it, it is possible to create an automatic execution customer experience, where the firm solves the need of the customer proactively.

We want to reiterate that automatic execution should not be seen as the most desirable customer experience for every transaction. Customers differ in how much agency they prefer, and for some transactions the risk of getting it wrong with automatic execution outweighs the benefits. While technologists might see automatic execution as nirvana, good old-fashioned customer understanding is necessary to offer the most relevant experience to your customers, which may require you to create a range of connected customer experiences.

5

Repeat

Building Customer Relationships to Create Competitive Advantage

---

Henry Ford's quip on color choices for his legendary Model T illustrates the trade-off between willingness-to-pay and production efficiency: "Any customer can have a car painted any color that he wants so long as it is black." Ford had no proclivity for black paint. His first car, the Model A, came in red, and the Model F was primarily sold in green. Instead, Ford's cookie-cutter strategy and one-size-fits-all paradigm came to favor production efficiency over customization.

That same production efficiency preference extends beyond manufacturing to the world of education, an industry that in the United States employs some 3 million K–12 teachers and another 1.7 million faculty in postsecondary education. Curricula are standardized. In France, the Ministry of Education dictates what every student will learn each day. In England, it's called the National Curriculum. Curricula are the assembly instructions for the production facilities of education.

Having a curriculum is not a bad thing. It holds teachers accountable and helps students achieve predefined learning objectives. It also helps coordinate across courses and schools, and fosters the sharing of best practices. Nevertheless, a standardized curriculum wastes an enormous opportunity for customization. Students have varying motivation, prior knowledge, and maybe talent. A student going through K–12 education will interact with some one hundred teachers and counselors, each one following a different piece of the curriculum.

What is the alternative, short of providing each kid with a set of private teachers? Fortunately, there are alternatives made possible by connected strategies. Consider the following three examples.

In 2006, Salman Khan, an MIT-trained computer scientist with an MBA from Harvard, launched a revolution in K–12 education called Khan Academy. Khan, at that time employed by a hedge fund in Boston, had been tutoring his cousin Nadia, who was struggling with basic math problems and couldn't get placed in a more advanced class. In addition to phoning her, Khan used a technology called Yahoo Doodle to scribble on a virtual notepad that he shared with Nadia through the internet. As this tutoring proved to be effective, he started teaching her siblings. By 2006, word about his remarkable teaching skills got out and Khan started to upload on YouTube simple videos of himself scribbling notes with some voice annotation. It was the foundation for what became the nonprofit Khan Academy. Ten years later, Khan Academy has over one hundred employees and has amassed twenty thousand videos used by fifty million students and schools around the world.

As a second example, consider the recent development of smart textbooks for college students, discussed in chapter 1. For many generations of students, the only touch point between student and publisher was the retail store, either brick-and-mortar or online. Thanks to online books, a digital connection is now made with the student every time the book is opened. What is the benefit of this? First, publishers (and professors) can track learning activities such as reading or homework preparation. Not only is such automated grading more efficient for the college, it also provides immediate feedback to the student. Immediate feedback is essential for learning. Rather than waiting for the final exam and getting a C because of insufficient preparation, the student knows where he or she stands with respect to the learning objectives of the course and thus can take any necessary corrective actions quickly and without compromising the final grade. Mistakes are made early in the learning journey, and the smartbook guides the student by showing recorded videos of solutions to similar problems or by redirecting the student to the relevant chapters. When the student is ready to move on, the learning activities can be completed in thirty minutes. If, however, the student struggles, the book is patient and guides the student through more hours of learning. Second, the learning activities of the student population create data, often referred to as metadata. Professors can use such metadata to decide what topics need further clarification in the upcoming class sessions. Authors and publishers can use the metadata in deciding what to write and publish next.

Finally, consider the example of Lynda.com, a company that was acquired by LinkedIn for $1.5 billion. Started by Lynda Weinman, it offers video courses geared toward professional skills, such as software development, graphic design, and business. But learners at Lynda.com don't learn for the purpose of passing a test. Instead, they articulate career objectives by picking a learning path. These paths could be digital marketer, web developer, or IT security specialist. Lynda.com then provides a bundle of video instructions, practice assignments, certification, and career management. Learners use Lynda.com not by simply asking it for one course (such as Essentials in JavaScript) but by entrusting the site with broader career objectives ("Make me a web developer"). At the level of the course, and even more so at the level of a single video lecture, Lynda.com competes with online courses and free YouTube videos. But having been entrusted with its learners' career ambitions, Lynda.com has secured a position of ongoing personal connection and trust.

This chapter explores the repeat dimension of connected customer relationships. Fundamentally, the repeat dimension strengthens the other three design dimensions that are involved in creating a connected customer relationship: recognize, request, and respond. As you have likely guessed by now, we will be drawing examples from the edtech (educational technology) industry, though many other cases are discussed as well.

After briefly describing how new technologies have shifted the frontier in the world of education, we will introduce a four-level framework of customization. This framework outlines how repeated customer interactions can be used to shift the frontier defined by willingness-to-pay and fulfillment costs. The four levels are:

1. Create unified customer experiences across episodes.

2. Improve customization based on past interactions.

3. Learn at the population level to enhance product offerings.

4. Become a trusted partner to the customer.

A Shift in the Efficiency Frontier in Education

Earlier, we discussed the concept of the efficiency frontier. Firms face a trade-off between lowering their costs and increasing their customers' willingness-to-pay by providing them with better or more convenient products or services. In chapter 2, we saw how Blue Apron and Uber shifted this frontier in their respective industries to raise the customers' willingness-to-pay while paradoxically lowering costs.

What does the efficiency frontier look like in education? If you go back in history, private teachers educating the aristocratic elite through one-on-one tutoring seems to be one of the earliest forms of formal education. The power of one-on-one instruction is obvious: The teacher can spend all her effort and attention on the unique needs of one student. Content and speed of instruction can be customized. If the private teacher comes to the student's home (or castle), convenience for the student is also maximized. But it is very costly and inefficient from the perspective of the teacher. In the modern era, the teacher would

FIGURE 5-1

The traditional efficiency frontier in education

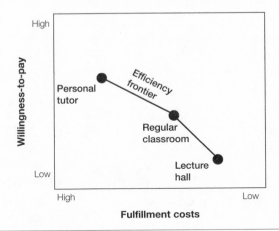

much rather explain the concept of quadratic equations once to a class of thirty rather than having thirty students individually try to learn this skill during her office hours. From an efficiency point of view, it would be even better to give the lecture of quadratic equations in a huge lecture hall, as is commonly done in introductory courses at universities. At the same time, the student's happiness and effectiveness of learning is reduced. We can see this trade-off in figure 5-1. (Recall that willingness-to-pay reflects the benefits that a student receives, not the price the student is actually paying.)

Private teaching is what Khan provided to his cousin Nadia, much to her benefit. This was only possible because of the love and empathy of her uncle. Let's do a quick back-of-the-envelope calculation of the costs of having this particular private tutor. As a Harvard MBA working at a hedge fund, Khan likely made somewhere between $500,000 and $1 million per year. Even if he worked many hours per day and rarely took a day off, his average hourly compensation must have been between $300 and $400 per hour. But in an educational setting, a cost of over $300 per hour per student is not a scalable model.

Now let us turn our attention to quality, or how much a student benefits from a particular way education is delivered. Most likely, these benefits are a function of the following factors:

The quality of the instructor

The customization of the content relative to the student's interests, career ambitions, and learning style

The degree to which the speed of instruction is customized to the ability of the student

The convenience of the educational service in terms of the timing and location of classes

Because of economies of scale, teaching classes of one hundred is much more efficient than teaching classes of ten. This is the reason why educational institutions have long discussions about faculty-student ratios. But the trade-off is that, in classes of one hundred, it is hard to customize either the content or the speed of instruction, not to mention the timing and location of the class.

This brings us back to the idea of shifting the frontier. When the instructional videos produced by Khan were uploaded on YouTube, the cost of production, including his time, was amortized by many more students. On YouTube, EdX, or Coursera, many video lectures have been watched by tens of thousands. Even if we factor in the costs of production, including expenses such as video editing and production (which make it more expensive than simply lecturing in front of students), the cost per lecture per student is reduced to pennies.

But what about the student benefits derived from such video lectures? Aren't they just as bad as Ford's slogan, "Any customer can have a car painted any color that he wants so long as it is black"? The answer has been the biggest surprise to all those active in online teaching: no. To understand why, let's go back to the drivers of student benefits mentioned earlier.

First, consider the quality of the instructor. Hundreds of years ago, those seeking entertainment and amusement would go to the local market to watch a clown or acrobat. Clowns were not making fortunes, but there was enough demand for this type of labor that pretty much any town would bring enough business to support its own clown. The clown profession, however, changed abruptly with the introduction of film technology. In times of movie theaters, the production of films was centralized, reducing the cost per laugh and limiting the demand for clown labor. For clowns, this was a sad story, but not so for the audience. Because the best clowns would star in the movies, the audience now could watch those who were really funny. Teachers and clowns have more in common than most in our profession would like to admit. With 3 million teachers in the United States, we have about 250,000 teachers per grade level. If we further break this up by subject, we end up with some 50,000 math teachers in eighth grade. Every one of these 50,000 teachers will explain the concept of quadratic equations each year. Most of them will do it well. Nevertheless, the idea of watching the very best teacher on video is increasingly appealing to students and parents alike.

Who teaches where has long limited the ability of schools to offer a wide range of topics. For instance, which foreign language you learned in elementary school, if you had the privilege of learning one at all, depended on which school you attended. Few elementary schools have the resources to teach French, Spanish, Mandarin, German, and Hebrew. In contrast, platforms like Rosetta Stone that serve a national and even global market have scale. They are thus in a much better position to provide students with the language instruction they desire.

The award for Most Popular Customization Tool in Online Education (we made that up) should be given to the pause button on the video player. In a lecture hall with one hundred students, there are only so many times a teacher or professor can pause and repeat herself. Online, there are no limits. If a student is distracted, the content is difficult, or the explanation of the professor is unclear, all it takes is a click and the video is paused, allowing the student to reflect and replay. As experienced online teachers, we also learned from our students that the second-most-popular tool is that for adjusting the video speed. Apparently, when watched at 1.5 times the normal speed, some of our most boring lectures become tolerable.

Finally, there is the effect of convenience. The new generation of learners, whom we as professors now teach, grew up with online devices and are accustomed to the "anytime, anywhere" paradigm of our society. From Khan Academy to smartbooks and from Rosetta Stone to Lynda.com, convenience is a key need and expectation of many users that we as educators might not welcome but must embrace.

Before we continue, a clarifying comment is in order. Being both parents and experienced online teachers, we by no means want to imply that kids should be educated by video instruction alone. Teachers will always play an important role in education. Nevertheless, technology has changed the way education is organized and has shifted the frontier, leading to bigger student benefits at lower costs. The following sections dive into greater detail on how the repeat dimension especially can shift the frontier in education and other industries.

Create Unified Customer Experiences across Episodes: Strengthen "Recognize"

Up to this point, we have discussed customer experiences with firms by looking at one episode or transaction at a time. But the greatest potential of connected strategies lies in creating deep, ongoing *relationships* with customers that weave together multiple experiences. The repeat dimension is thus fundamental in transforming stand-alone experiences into relationships. The first step to achieve this goal might sound trivial, but it is essential and turns out to be quite difficult: you need to be able to identify the customer and treat him or her as the same person whenever you interact, regardless of when and where this interaction takes place. Only if you keep track of your customers will you be able to learn more about them—that is, improve the recognize dimension of connected customer relationships.

Such a customer-centric view is remarkably uncommon. For instance, in the world of education, students traditionally interact with the school or university one course at a time, and it is up to the student to stitch together a coherent experience. A connected strategy approach, in contrast, focuses on the learner, not the course. This allows the aggregation of otherwise disjointed learning experiences into one unified learning journey. Teachers and counselors have access to the data of past student performance, and no student falls through the cracks, which increases the quality of the instruction. Costs also come down at the same time by saving teacher and counselor time that is otherwise spent trying to make sense of poor student performance that could have been predicted (and avoided) much earlier.

Similarly, in the world of health care, most of us have experienced the annoyance of checking in at a physician's office. How many times must we as patients provide our medical history, our allergies, and our insurance information? Wouldn't it be nice if, when our sleep pattern suddenly becomes abnormal, our physician is put into the loop? Chances are that if we are Apple Watch users, Apple now knows more about our health than our doctor, for whom we are patients when sitting in the exam room but strangers when we are not.

The problem of orchestrating all interactions and weaving them together into a unified customer experience is harder than it first appears. The reason lies in the fact that many companies now interact with each customer through multiple channels. This creates at least two problems.

The first is a technological one. Complex businesses with multiple product lines often do not use a single database or IT infrastructure. For instance, when a firm interacts with a customer through both traditional brick-and-mortar retailing and online channels (omni-channel retailing), it is quite problematic to track that customer across all her interactions with the firm's various touch points.

This leads us to the second problem, an organizational one. The reason for the multiple IT systems is often historical. Different business units develop their own processes and systems, a problem that is exacerbated when divisions are added through mergers and acquisitions. Moreover, these units often fight for internal resources or compete for status and career slots. So, when a customer who was well advised by an employee in a retail store ends up buying a product from the online branch of the same retailer, the store manager might view him as a customer who was lost to another unit. Similarly, consider Disney. To generate the amazing guest experiences that we described previously, Disney had to overcome exactly these challenges:

> The data related to a given customer was scattered among (for example) the Disney video games the customer had on her PlayStation, the retail store at which she purchased the last piece of Disney apparel, the Disney movie she saw on Netflix, the Disney theme park she visited last year, and the Disney Hotel she stayed in. Integrating this into a single customer relationship is not easy, but without integration, how could Bill, who was acting as Captain Jack Sparrow in Disney's Anaheim park, remember that little Sydney had seen Bill's colleague, François, in Disney's Paris park last year?

> Even though they are part of the same company, theme parks have to be profitable, and so do feature films. To move from a product-line-(channel)-based view of the world to one that puts the customer in the center of all transactions requires strong vision and leadership support from above. Traditionally, it was the customer who had to navigate Disney's organizational chart in order to stitch together a seamless experience. As part of the MagicBand implementation, organizational changes had to be executed as well.

But Disney did it, and you already know the results from chapter 1. Not only did the guest experience improve, but in many cases, the costs also dropped. Where it was once necessary to manually weave together transactions across channels and time when handling special customer requests or complaints, now a seamless customer experience can be delivered with high efficiency.

Improve Customization Based on Past Interactions: Strengthen "Request"

While the first level of customization is all about keeping track of and getting to know the customer across individual transactions with the firm, the second level is about turning this information into actionable knowledge. The firm needs to use the information about a customer's needs to translate it into a specific request for an appropriate product or service. To understand which product or service is most appropriate, the firm needs to understand which willingness-to-pay drivers are particularly important for a given customer.

In the last chapter, we introduced the concept of the customer journey (see figure 4-1). At each step of this customer journey are a number of possible willingness-to-pay drivers. Understanding these drivers is essential to customizing the customer's experience. (We will guide you through this process in the next workshop chapter.) Most importantly, what the customer journey highlights is that your customer's willingness-to-pay is driven not only by the product or service itself (the "what") but also by how a customer interacts with you and how the customer can access your products.

For instance, convenience of access has become an ever more important element of customization. Again, the world of education provides an illustration. In the old world of brick-and-mortar education, physical campuses and class schedules created a rigid delivery system. In today's world, "anywhere and anytime" has become the mantra of online education, especially in the market of educating busy professionals. Thus, physical campuses and fixed class schedules are inconveniences that negatively impact the customer willingness-to-pay.

Customization goes beyond the ability to access content anytime the learner wants. While it sounds great to have access to tens of thousands of educational videos around the clock, the options can be overwhelming. Maybe a learner wants to become a web developer. Coursera, EdX, and even YouTube have plenty of video material that the learner would benefit from. But where to start? As mentioned earlier, Lynda.com bundles videos so that they collectively correspond to a career track. It takes the expressed need of the learner ("I want to become a web developer") and turns it into a solution ("Take JavaScript first, then take a course on interface design," etc.). This is the idea of curation leading to customization. The smartbook at McGraw-Hill takes customization one step further still. Beyond reacting to explicitly expressed user needs, it also infers customer needs based on past interactions. Past reading and test-taking behaviors are analyzed and used for future curation.

This is the skill that Amazon has mastered so well. By observing our past browsing and buying behavior, Amazon is able to infer our needs. Moreover, it creates a virtuous cycle. The more a firm engages in business with somebody, the more it learns about the customer and the better it is able to customize future offerings. The better the firm customizes its offerings, the more delighted the customer becomes, bringing the customer back again and again, creating even more information for the firm. At some point, the customization becomes so good that customers get locked in and stop taking their business to competitors. Recent data shows that Amazon has more than 40 percent market share of online retail. This feedback loop is visualized in figure 5-2. Given one customer, a firm learns more and more about what that one customer needs. This creates a positive feedback loop: recognize, request, respond, and repeat, then recognize, request, and respond even better, and so on.

FIGURE 5-2

Learning at the level of the individual customer

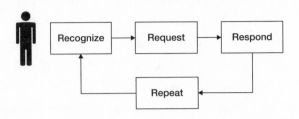

Learn at the Population Level to Enhance
Product Offerings: Strengthen "Respond"

We recently worked with a telecommunications executive who shared the following story. He was at the checkout counter at a large home-improvement store. The cashier asked for his zip code, to which he responded, "I will tell you my zip code if you give me a 5 percent discount. In fact, if you give me 10 percent off, I will tell you the street I live on." The clerk called the manager, who took the deal!

It is said that, in this connected world, customers pay not just with their wallets but also with their data. We will discuss this theme further in chapter 8. For now, let us simply observe that knowing your zip code and your street address does not just allow the store to serve you better; the store can also transfer this knowledge to better serve other customers like you. Conversely, it can use the data on customers like you to help in predicting what you might need. This is the first advantage that comes from population-level learning. The firm can move beyond using an individual's data to help that person by using aggregate data to make customized suggestions or decisions for each of its customers.

Population-level data makes even more powerful learning possible. By learning about its customer population, the firm can create a better product or service offering. After all, what good is it to have a deep understanding of your customers' needs if you don't have the products or services available to satisfy those needs? True customization requires not only understanding the customer deeply but also having the right product and services available. Thus, level 3 of customization is fundamentally about strengthening your ability to respond.

First, consider examples from the world of education. Learning analytics is emerging as a hot new field. If we can predict which students are likely to struggle in a course, we can take corrective action before problems occur. Teachers can learn where individual students or entire classes are likely to struggle, allowing them to proactively alter what they offer in their courses. The same can be true for authors like us. If, for example, we knew that learners love the worksheets in chapter 3 but rarely use the ones in chapter 10, we could improve this book. In fact, we hope to achieve exactly this through our website, connected -strategy.com.

A parallel trend is playing out in medicine. Under the label of personalized medicine or precision medicine, health care companies mine genomics data in the hope of finding predictive patterns for who will develop Alzheimer's or cancer, among a range of illnesses. For example, the genetic testing firm 23andMe is establishing itself as a valuable partner to biotech companies as it amasses the genetic profiles of millions of people.

As a firm learns more about its customers, it can also broaden the set of customer experiences that it creates. Consider Square, a financial services provider founded in 2009. Square started out by providing small businesses with a lower-cost option for accepting payments via credit cards. Through its Square readers (small electronic devices for swiping cards), Square helps its clients to improve their respond-to-desire strategies. Over time, as Square learned more about the needs of its clients, it created curated offerings that included new features such as tailored dashboards providing information about the end customers and new services such as payroll systems. The information contained in the Square system also allows small businesses to provide more curated offerings to their customers—for instance, via targeted advertising. Lastly, Square has started to offer an automatic execution experience by automatically issuing lines of credit in real time based on the merchant's cash flow.

As these examples illustrate, population-level learning allows firms to refine their product portfolio in two different ways. First, learning about demand allows a firm to better choose which products it should carry. The second type of portfolio adjustment is more radical. As a firm learns more about its customers, it might get deeper insights into them than any of its suppliers have. These insights might then allow the firm to backward integrate and produce (or direct suppliers to produce) brand-new products. Consider Zalando, one of the largest German online fashion retailers. Zalando started as a copycat of Zappos, the largest online retailer of footwear in the United States. Zalando initially focused on providing a respond-to-desire customer experience. Over time, as Zalando learned more about its customers, and customers were willing to share personal information and fashion preferences, Zalando was able to add curated offering activities, matching individual customers with selected items that are presented to them through the company's website. Eventually, Zalando was also able to use the data it gathered to start a private-label brand. From searches on its website, Zalando had customer data for which price points and product categories customers rarely used a "brand" filter. Zalando realized that for these products, its customers didn't care much about the brand name. Therefore, Zalando started to offer its own products in these categories. (See the sidebar for another example of repeat in action.)

Again, we can observe a virtuous cycle, a positive feedback loop. The larger the set of customers a firm serves, the more information it can gather to fine-tune its existing product portfolio through better assortment or the creation of new products. The better its product portfolio, the more likely it is that it can find a good match between the needs of a customer and its product offering. This good match, in turn, leads to customer satisfaction and expands the customer pool, again creating more data.

ASTHMA INHALERS IN THE REPEAT LOOP

Nonadherence to medication is a key cost driver in health care systems around the world. In the United States alone, $100 billion to $300 billion of avoidable health care cost has been attributed to nonadherence. It's particularly problematic for long-term, chronic diseases where patients are not always symptomatic. For instance, the World Health Organization has estimated that almost half of all prescribed medication for asthma is not being used. This can lead to costly emergency room visits, hospitalizations, and emotional trauma, as many parents can attest if they have endured late-night hospital visits for a child with an acute asthma attack. No wonder many firms are trying to reduce these costs, while providing more value to patients, via connected strategies. Consider the SmartInhaler developed by New Zealand–based Adherium. The SmartInhaler is a Bluetooth-enabled sensor that wraps around the patient's existing inhaler. The device sends information via an app to the patient, or parents, and health care professionals to track medication adherence. After learning the average usage pattern of a patient, the app also sends the patient reminders or alerts if a dosage is missed. Over time, the app learns more about a patient and can start predicting when asthma may strike, allowing the patient to preempt attacks. The device contains a range of sensors that allow feedback to the user not only about whether the medication has been taken but also about whether the inhaler has been primed correctly and pointed accurately inside the mouth to deliver the full dosage to the respiratory system. Given its information on its population of users, Adherium has been able to provide valuable feedback to AstraZeneca, the maker of the inhaler, to help it redesign the inhaler so that it is more likely to be used correctly. In this case, we can see the repeat dimension in full play. Over time, the app learns more and more about a particular patient, allowing it to improve its coach behavior. Likewise, learning at the population level allows the app to improve its analytics with respect to predicting asthma attacks, thus improving the device over time.

FIGURE 5-3

Population-level learning

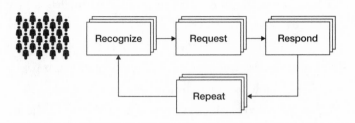

Figure 5-3 illustrates this positive feedback loop. Unlike figure 5-2, which was all about finding what is best for one particular customer in order to provide a better curation, metadata enables learning about many customers.

Become a Trusted Partner to the Customer: Recognizing Deeper Needs

As a firm learns more about its customers, it also has the opportunity to move from addressing one or more narrow needs to focusing on more fundamental ones. A narrow need is to learn about compound interest rates. A more fundamental need is to be able to ascertain the value of an investment. More fundamental still is the desire to become an investment adviser. When a learner entrusts her career dreams to Lynda.com, customization can be done at a whole new level because Lynda.com can take a more active role in the connected relationship. Beyond curation, the firm might nudge the learner to keep on top of her homework (coach behavior) or even automatically sign her up for an important job fair.

The distinction between narrow needs and more fundamental ones is also relevant in the field of health care. If a patient feels some heart palpitations, the narrow need is to talk to a cardiologist. More broadly, what this patient wants is to have the health care provider deal with her cardiac problems. Actually, what the patient really wants is for her health care team to provide the right health care when needed. Most fundamentally, what the person wants is for her health care team to keep her healthy. Thus, we can identify a hierarchy of needs in which the current request is an expression of a higher-level, more general need. The promise of connected strategy is that through repeated interaction, a firm is able to move up this hierarchy of needs and embed each user experience in a deeper relationship between the firm and the customer. In doing so, firms can address more fundamental drivers of customer value, increasing a firm's value proposition.

A helpful approach to discover such deeper relationships by addressing more fundamental needs is the why-how ladder. Figure 5-4 provides an illustration for our cardiology example. Each of the boxes in the why-how ladder corresponds to a specific problem definition. Problem definitions at the bottom of the ladder are more focused, addressing *how* a need could be fulfilled. We climb up the ladder by asking *why*. Why is that problem relevant in the first place? Why would it be good to fulfill this customer need?

Going up the why-how ladder accomplishes two goals. First, it aligns the search for solutions with what the customer really cares about. Again, patients don't really care that much about their cardiologist; they just want to make sure that their heart is in good shape and, even more broadly, that they are healthy. That's the most relevant problem for the patient, and whoever provides the solution to this problem is likely to win the competition to serve this patient.

Second, this understanding opens up alternative solution approaches: solving the problem of providing easy access to a busy cardiologist is hard. At this level in the why-how ladder, our solution space is limited to finding more cardiologists and making them work faster or longer hours. But as we go up to the problem of keeping a patient's heart healthy, there exist many alternative solutions, ranging from changing exercise routines and nutrition to reinforcing medication adherence. For every dollar that we spend, we might be able to improve patient cardiac health by a lot more if we invest in methods to rein-

FIGURE 5-4

The why-how ladder for cardiology problems

In the eyes of the customer, the purpose of the relationship with our firm is to ...

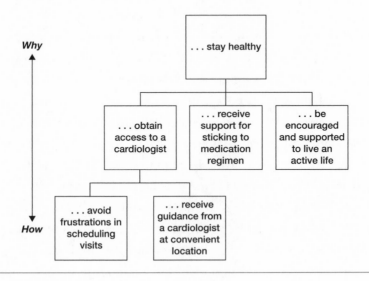

force medication adherence or lifestyle management. This improves efficiency.

Researchers at the University of Pennsylvania conducted clinical studies on cardiac health and found some interesting results. Prior studies had shown that many of those discharged from the hospital after being treated for major cardiac problems were not willing or able to stay on their medication for longer than six months. Using pill bottles connected to the internet, the Penn research team could quickly detect when patients forgot to take their medications. By automatically hovering over the patient this way, deviations can be detected early and patients can be engaged and trained to form healthy behaviors. The researchers used small financial incentives and peer pressure built through social media to coach patient behavior, nudging them to stay on their medications and lead a healthier lifestyle.

Clearly, a firm needs to earn the trust of the customer before it is permitted to manage a more fundamental need. This is why we put this at the fourth and highest level of how the repeat dimension transforms customer experiences into customized, connected customer relationships. There is an interesting circularity here: Only if you have a deep connection with a customer—relying on intensive data exchange—will you be able to address more fundamental needs. At the same time, unless you are able to address more fundamental needs, customers likely won't want to engage in a deep relationship with your firm in the first place. Deep, embedded connections can be intrusive; customers will have serious and justifiable privacy concerns. Unless the value delivered to the customer is high, customers will not want to engage deeply or may feel that their data is being exploited without their consent. Thus, you won't be able to jump straight to level 4 of customization. Level 4 is achieved in stages. The customer allows you access to a certain amount of data. Once you have proven to the customer that this data enables you to make the customer's life better, the customer might grant you access to the next slice of data.

As you can see, the repeat dimension at all four levels of customization helps a firm to shift the efficiency frontier. A better understanding of customer needs, a better ability to translate those needs into specific product requests, and a better assortment of products that fulfill those needs precisely all increase the willingness-to-pay of customers. At the same time, a better understanding of demand allows the firm to avoid inefficiencies. Table 5-1 summarizes the four levels and their impacts on willingness-to-pay and fulfillment costs.

The Importance of the Repeat Dimension for Creating Sustainable Competitive Advantage

The repeat dimension of connected strategies moves the relationship from episodic transactions to a continuous relationship. Once the individual transactions are woven together into a customer-centric, unified experience (level 1), a firm has set itself up to serve its customers better and more efficiently. This improvement, and the associated shift

TABLE 5-1

The four levels of customization created by the repeat dimension

Level	Impact on willingness-to-pay	Impact on cost
Level 1: Create unified customer experiences across episodes	Customer is treated as one person across channels and transactions	Avoids manual weaving together of experiences
Level 2: Improve customization based on past interactions	Ability to identify offerings that address the willingness-to-pay drivers most important to the particular customer	Avoids costly iterations in case of failure to fulfill the need
Level 3: Learn at the population level to enhance product offerings	Higher-valued offerings based on inferring customer needs	Data-driven approach to innovation
Level 4: Become a trusted partner to the customer	Addressing more fundamental needs allows for alternative solutions and early interventions	More efficient use of resources, as solution space is broadened

in the efficiency frontier, is made possible by two learning mechanisms, summarized in figure 5-5.

The first mechanism plays out at the level of the individual. As a firm engages in more interactions with that customer, the firm better understands the customer's current needs and what products or services would best fulfill those needs. This is level 2 in our framework. For respond-to-desire customer experiences, the firm also can help the customer in understanding and expressing his or her needs more precisely. Thus, the first mechanism across levels 1 and 2 strengthens the dimensions of recognize and request.

While it is wonderful to have a deep understanding of your customer needs, this information is not very valuable unless you have the products or services available to satisfy those particular needs. The second learning mechanism operates at the level of the population (or the segment) by analyzing metadata. This learning creates a feedback into the assortment of products or even creation of new products in the first place: "Given what we have learned about customers of varying types, what would be the optimal assortment to carry or products

FIGURE 5-5

The positive learning feedback loops created by the repeat dimension

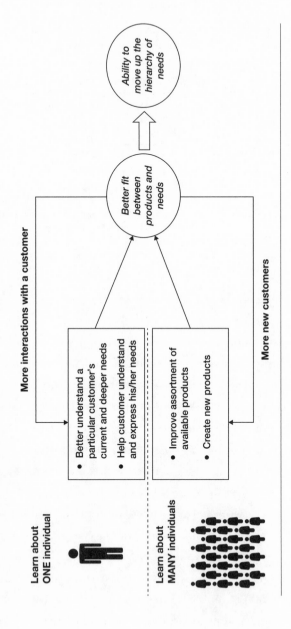

to create?" In short, this learning mechanism improves the dimension of respond. This is level 3 in our framework.

Together these learning mechanisms allow a firm to enhance the personalization of its offering. The firm can create a better fit between the needs of the customer and the product (or service) that responds to this need. The more Netflix knows about Samantha's viewing habits, the kinds of films her friends are tweeting about, or perhaps her upcoming vacation plans, the better Netflix is able to personalize her viewing recommendations ("Flying to Italy? Watch *Tuscan Wedding* to get into the mood!"). At the same time, as Netflix learns more about entire customer segments, it can optimize not only what kind of content to license but also what kind of content to produce. The data Netflix is able to gather from its more than one hundred million subscribers worldwide has allowed it to create more than twenty-seven thousand genres, including genres such as 20th-Century Period Pieces Based on Classic Literature, Absurd Opposites-Attract Comedies, and Biographical Fashion Documentaries. This fine-grained categorization, combined with viewer feedback and observed behavior at both the individual and the population levels, gives Netflix deeper insights into its audience than any movie studio could ever hope for.

Eventually, Netflix or other firms will be able to use this information to move up the hierarchy of needs of their customers and achieve level 4 of customization. Yes, a customer wants to watch a movie at certain times, but the deeper need might be entertainment. Once a firm understands a customer deeply, not only can it suggest movies, but it can also arrange for tickets to live concerts, automatically record sporting events, and play the customer's favorite music in her house and car.

What makes the repeat dimension so powerful is that it involves positive feedback effects that over time can create a tremendous, sustainable competitive advantage for a firm. As we see in figure 5-5, the tight fit between customer needs and available products—that is, the high degree of personalization—leads to more value created by the firm, either in the form of higher willingness-to-pay by the customer or by higher efficiency. This allows the firm to provide more value to current customers, creating more future interactions with these customers, which increases the individual-level learning (the top feedback loop in figure 5-5). At the same time, the increased value allows the firm to attract new customers, thereby enhancing the population-level learning (the bottom feedback loop in figure 5-5). With more learning at the individual and population levels, the firm continuously improves the recognize, request, and respond dimensions, creating ever-increasing degrees of personalization. It is a process that feeds on itself and can allow a firm that gets ahead of its competitors to continue to expand its competitive advantage.

Moreover, as a firm is able to improve its knowledge about its customers' needs and its ability to service these needs, it has the ability to move up the hierarchy of needs of its customers. Once the firm has transformed a series of customer experiences into a true relationship, customers will be much less likely switch to other firms. Firms with established connected relationships with their customers do not have to compete transaction by transaction for the business of their customers because they have created an effective lock-in. To woo customers away, competitors have to work much harder than simply offering an occasional better deal. As a matter of fact, if you are able to reach the status of a trusted partner, customers are quite likely to become advocates for you, telling their friends about the great service they receive.

In our foregoing discussion, we stress the *learning* feedback loops of the repeat dimension, as they have been most underappreciated and underexploited in our experience. As a firm gains more customers, three better-known positive feedback loops can also arise that will further strengthen a firm's competitive advantage.

First, as a firm attracts more customers, it will enjoy economies of scale: fixed investments can be spread over a larger customer base. For instance, Amazon's investments in recommendation engines, website design, and technological improvements in Alexa can all be spread over its millions of customers, giving it a cost advantage over firms with fewer customers. Economies of scale allow a firm either to offer increasingly better products without having to raise prices, or to lower its prices to its customers—or both.

Second, as firms attract more customers, network effects can arise. Network effects exist when the willingness-to-pay of customers increases with the number of other users. For instance, the more that people use Facebook, the more likely it is that the next user will pick Facebook because all his or her friends are on this platform. That, in turn, increases Facebook's user base even more.

The third positive feedback loop is a two-sided network effect that exists when more participants on one side of a transaction increase the value for the participants on the other side of the transaction, and vice versa. For instance, the more customers Apple is able to attract to its App Store, the higher the incentives for software developers to write apps and post them in the store. At the same time, the more apps that are available, the more customers are attracted. Likewise, the more that customers use a ride-hailing service like Lyft, the easier it is to attract new drivers; and conversely, the more drivers a ride-hailing service has, the shorter the wait times and the more likely that a customer will choose this particular service. All of these positive feedback effects create ever-increasing advantages as a firm grows faster than its competitors.

As we noted at the end of chapter 2, creating new and superior connected customer experiences is only the first step of building a successful connected strategy. If you can utilize technological advances to create a better customer experience, so can your competitors. But if you can go through the recognize-request-respond loop more often and learn more than your competitors each time you repeat the cycle, you can indeed create a competitive advantage that is sustainable. While all the firms we use as examples in this book have been innovative in creating new connected customer experiences, only those that are able to utilize the repeat dimension thoroughly, and create and exploit the various positive feedback loops, will be successful in the long run.

The Data Trust Challenge of Connected Strategies

As we depicted in figure 5-5, two feedback loops are at the heart of a connected relationship: by repeatedly having interactions with one particular customer, the firm is able to better and more efficiently serve that customer; and by obtaining information about many customers, the firm can better position itself for the future.

The resulting competitive advantage can lead to market share and profitability. That is great for the firm, but is it great for the customer? As firms perfect their service to a particular customer (upper part of figure 5-5), two investments have to be made. One investment is the data-collection and analysis effort of the firm—listening carefully to the needs of the customer and learning from one episode to the next. The second investment is made by the customer, who has to share information with the firm, be it actively by answering questions and expressing preferences ("Siri, wake me up every Monday at seven o'clock and order my coffee from Starbucks"), or by permitting passive monitoring by the firm (e.g., allowing a fitness app to track sleep patterns). So, the value that is inherent in a successful connected relationship, the force that allows the firm to shift the efficiency frontier, is coproduced by firm and customer.

This coproduction concept is not just a matter of semantics; it raises customers' expectations of how much value they should receive from the relationship. Unless customers think they are getting a fair share, they may want to quit the relationship and you will never reach level 4, becoming a trusted partner.

More generally, firms will not be able to maintain the repeat dimension if they lose the trust of their customers. Because a rich information flow from the customer to the firm is central to a successful connected strategy, data privacy, data security, and transparent data use are absolutely essential.

Both the regulatory space and user attitudes toward privacy are likely to change over time. As a result, guidelines will evolve. Still, privacy guidelines from the Organisation for Economic Co-operation and Development and the European Union's General Data Protection Regulation are good starting points for your considerations. To build a connected strategy, you will have to have policies that address these guidelines, including the following:

1. *Collection consent:* Whenever you collect data, it should only be with the knowledge and consent of the individuals you collect it from. Customers should have the right to withdraw this consent subsequently.

2. *Data quality:* It is your responsibility to keep data accurate and up to date for the purposes for which it is to be used.

3. *Purpose:* You need to state clearly the purpose for which data is collected before collection starts, and that purpose should not be changed unless you notify the customer.

4. *Nondisclosure:* The data you have collected should not be disclosed or made available to others except with the consent of the individuals it is collected from.

5. *Safety and breach notification:* It is your responsibility to protect data against unauthorized access or disclosure. Should a breach occur, it is your responsibility to notify your customers in a timely manner (within a few days).

6. *Openness:* Your customers should be able to easily understand who is collecting data and for what purposes.

7. *Access:* Your customers should have the right to access the data that you have collected and to have corrections made if the data is not accurate.

8. *Data portability:* Your customers should have the right to receive their data in a commonly used and machine-readable format and have the right to transmit this data to another firm.

9. ***Data erasure:*** Your customers should have the right to have their data erased and to stop further dissemination of their data.

10. ***Accountability:*** You must commit to being held accountable for following the foregoing principles.

Four Levels of Customization to Become a Trusted Partner

The repeat element in a connected customer relationship can often raise a chicken-and-egg problem:

> To provide a customer the level of customization that fulfills her deepest needs requires a strong connection, including large amounts of data from prior interactions.

> But to obtain the permission of the customer to collect large amounts of data requires that the firm is capable of providing a high level of customization and fulfilling the deepest needs of the customer.

How do you break into this seemingly closed loop? In this chapter, we have proposed that connected customer relationships get deeper and deeper over time by moving through four levels of customization:

> Level 1 is about creating a unified customer experience by weaving together previously unrelated episodes. Taking a customer-centric view, potentially across channels, can create efficiency gains by eliminating data reconciliation, is more convenient for the customer, and increases the amount of information that a firm has available on a particular customer.

> Level 2 uses the data from past interactions to improve customization and to learn which products or services are the most

important drivers of willingness-to-pay—that is, to determine what is truly requested by the customer.

Level 3 is about developing the capability of delivering on those drivers when and where desired by the customer. Responding to customer requests in an efficient manner requires the firm to aggregate information across many customers. This population-level learning improves its assortment of product or services.

Finally, level 4 corresponds to a move of the firm to tackle more fundamental needs, evolving from offering rental cars to becoming a mobility solution or from being a provider of accounting courses to becoming a source for business knowledge.

As a firm moves from one level to the next, it shifts the efficiency frontier and strengthens its relationship with its customers, thereby creating a competitive advantage. Even at the higher levels, the firm still needs to provide a more attractive option to its customers than its competitors, but it is freed from competing for every individual transaction. As we will see in chapter 8, this allows for revenue models that truly focus on long-term value creation.

6

Workshop 2

Building Connected
Customer Relationships

This workshop will systematically guide you in applying the content of the previous two chapters and assist you in building connected customer relationships. It has three parts.

In the first part, we help you diagnose the customer experiences that your firm currently provides. This will formalize some of what you already did in the workshop of chapter 3 by using the recognize, request, and respond dimensions of connected relationships discussed in chapter 4. More specifically, we will break up this diagnosis into three steps:

1. Map the current customer journey of one customer experience.

2. Identify customer willingness-to-pay drivers and pain points.

3. Capture the information flow for this customer experience.

The second part of the workshop helps you to look at the relationships you have with your customers across individual, episodic experiences.

Weaving together these episodes and learning from them to customize products or services for a particular customer was the first important element of the repeat dimension discussed in chapter 5. In chapter 5, we also discussed another element of the repeat dimension, the ability to learn at the level of the overall customer population.

The goal of a connected relationship is to move from a transactional relationship with your customer to becoming a trusted partner. Rather than only teaching finance, you help somebody make a career as an investment adviser; rather than only performing surgery, you support health; and rather than trading stocks, you help somebody save for retirement. This involves the following two additional diagnostic steps:

4. Identify the deeper needs of the customer.

5. Understand the current relationship with your customer across separate (repeated) customer experiences.

In the third part of the workshop, we help you turn the findings of your diagnosis into new ideas for creating connected relationships:

6. Identify new opportunities to reduce customer pain points and lower fulfillment costs.

7. Find ways to utilize information gathered from repeated interactions to improve the recognize-request-respond cycle.

Lastly, since trust is at the core of a connected customer relationship, we ask you to do the following:

8. Assess your data-protection policies to maintain trust with your customers.

Step 1: Map the Current Customer Journey of One Customer Experience

Just as you would begin any operational improvement project by mapping out your current process, we find it helpful to first map the typical journey your customers are taking when interacting with your organization. This journey starts with the emergence of a latent need, is followed by a customer's recognition of the need and request for something from your firm, and leads to your firm's responding to this request. If you have customer segments that experience very different customer journeys, sketch out one journey for each customer segment. Likewise, a customer may have different journeys with you (e.g., "buying insurance" and "dealing with a claim"). In that case, you again need to sketch out several journeys.

You can use worksheet 6-1 as a starting point. On this worksheet, we have outlined the stages of a typical customer journey as we explained them in chapter 4. For each stage, put an explicit description in each box. What does the customer actually do at each stage of the customer journey? Here are some questions you might want to consider for each stage as you fill out the worksheet:

Latent need

- What are the underlying needs that the customer wants to fulfill?

- What are the underlying problems that the customer is trying to solve?

Awareness of need

- When and how does the customer become aware of this need?

Search for options

- How does the customer search for options to fulfill the need?

- How many of the possible options are within the awareness of the customer?

Decide on options

- How does the customer decide on what option to choose?

- Who is involved in this decision? What advice does the customer seek?

Order and pay

- How does the customer order the option he or she has decided on?

- How is the customer being billed?

- How does the customer pay for the order?

Receive

- How long does it take for the product or service to arrive to the customer once the order is put in?

- Does the product or service come to the customer, or does the customer need to travel to the product or service?

Experience good or service

- How much effort is required before the customer can use the product?

- How does the customer use the product?

Postpurchase experience

- What are the postpurchase needs of the customer? Returns? Upgrades? Advice? Maintenance? Service? Replacement parts?

Customer journey

Why does the customer engage in the interaction?

How does the customer go about identifying, ordering, and paying for the desired product?

What products and services are provided to the customer?

Latent need	Awareness of need	Search for options	Decide on options	Order and pay	Receive	Experience good/ service	Post- purchase experience

Step 2: Identify Customer Willingness-to-Pay Drivers and Pain Points

In step 1, you documented the actual steps that a customer is going through on the customer journey. Now, let's focus on the willingness-to-pay drivers and the pain points that the customer encounters along this journey. If you have completed workshop 1, you already have a head start on this step.

Hopefully, the concepts we provided to you in the last two chapters allow your understanding of the willingness-to-pay drivers of your customers to be more encompassing and precise. For instance, in workshop 1 we asked you to think about inconvenience as a factor that could diminish willingness-to-pay. Now, with the tool of a customer journey, you can be more systematic in your analysis of where inconvenience can occur; for example, it is not only in receiving the product ("Where does the customer have to go and how long will it take to receive the product?") but also in the upfront part of the customer journey, such as in becoming aware of the need.

Many customers have a need but are simply not aware of it. If you just asked customers about their needs, they might not even be able to articulate them. In fact, a need that a customer is unable to express is the very definition of a latent need. We will dig deeper into this early stage of the customer journey in step 4 of this workshop, where we will present you with a tool to identify the deeper underlying needs that a customer might have. For now, we will start the customer journey with the stage of awareness of need.

Many of the examples we use throughout the book are from business-to-consumer settings. If you are in a business-to-business setting, you can ask yourself another overarching question: How can we help our customers create competitive advantage in their markets? Whatever you can do to help your customers become more efficient, or to increase the willingness-to-pay that their products create for *their* customers, will increase the willingness-to-pay that your customers have for your products.

Whether you are in a business-to-consumer or business-to-business setting, as you are thinking about the various pain points that your customer encounters, consider the following questions. Keep track of the willingness-to-pay drivers and pain points on worksheet 6-2. You will use them later in this workshop as ideas for possible opportunities for improvement.

Awareness of need

- How often is the customer entirely unaware of the need (e.g., the customer didn't realize the computer was at increased risk of being hacked)?

- How often does the customer in principle know of the need but fail to act on it at the current time and location (e.g., the customer knows that she should take a pill but simply forgets)?

- How often does a firm falsely remind the customer of a need that the customer actually didn't have or that was already fulfilled, creating customer dissatisfaction?

Search for options

- How do customers identify the options that could fulfill their needs? How much time do they spend on the search?

- Are customers generally aware of the most relevant options for them, or is the set of options simply too large and complex (e.g., finding the right tile for a bathroom renovation, or the right customer relationship management software)? In the normal search process, are new options being surfaced for the customer that the customer was not aware of, or is the customer mainly contemplating previously used solutions?

- How many options do you offer to the customer? What options are offered by competitors?

Decide on options

- What factors does the customer take into account in making a decision (beyond price)? How easy is it for the customer to assess each option for each of these factors?

- Does the customer use any outside help when deciding on an option (e.g., review sites, reputation scoring)?

- What role does trust in the provider play in the customer's decision making?

- How much effort does it take the customer to figure out which option would fulfill the need in the best way?

- How easy is it for the customer to understand how costly each option is (e.g., over the expected lifetime use)?

Order and pay

- Once the customer knows what she is looking for, how long does it take her to specify and order the product or service?

- How easy and convenient is it for the customer to specify where and when the product is to be delivered or the service is to be performed?

- How quickly is the customer being billed? How transparent are all charges that the customer incurs?

- How easy is it to pay? What forms of payment are accepted?

Receive

- Where will the customer receive the product? Does the customer need to pick it up, or will it be sent to the customer's desired location?

- How long does it take to receive the product after the customer has ordered it?

- What happens if the customer is not home when the product is delivered?

Experience good or service

- How much effort is required from the customer between receiving the product and deriving a benefit from it (time to unpack and install)?

- What are the technical features of the product that drive the customer's willingness-to-pay?

- What are the intangible product features, such as brand, image, and design, that drive the customer's willingness-to-pay?

- How good is the fit between the product or service and the customer's needs?

- Does the customer always get the product in the desired portion size?

- Does the product gain the customer access to complementary products and services (e.g., is it compatible with other third-party products, or older, already installed versions)?

Postpurchase experience

- How easy is it to reach customer support?

- Does the firm reduce the risk that the customer has to bear? How easy is it to return the product?

- What is the degree of postpurchase flexibility? The customer's needs may change after he or she has purchased the product. How easy is it to upgrade or downgrade the product?

Willingness-to-pay drivers and customer pain points at each stage of the customer journey

Why does the customer engage in the interaction?

How does the customer go about identifying, ordering, and paying for the desired product?

What products and services are provided to the customer?

Latent need	Awareness of need	Search for options	Decide on options	Order and pay	Receive	Experience good/ service	Post-purchase experience

Step 3: Capture the Information Flow for This Customer Experience

Now consider the information flow between your customers and your firm. Information could be an explicit request for a particular product by the customer ("I want product *xyz*"), but it could also be just information about the customer's state, be it the condition of the customer's printer or the customer's heart. Thus, information includes everything that the firm may be able to use to deduce and fulfill the needs of the customer.

For each of the steps you identified in the customer journey in worksheet 6-1, first describe the information that flows from the customer to you. Note your answers in the top row of worksheet 6-3. Then, for each information flow that you have identified, ask yourself the following questions:

- What or who triggers the information flow? Does the customer have to take the initiative?

- What is the frequency of the information flow? Does information flow in one batch or continuously?

- How rich is the information flow? Are you just exchanging a few bytes, or do you have access to high-bandwidth, multimedia information flow?

- How much customer effort is required? How long would it take for a typical customer to complete this step?

- Who is acting on the information at this step? Is it the customer who is primarily making the inference on how to use this information, or is it the firm that acts on this information? For instance, who is translating the customer need into a product or service provided by your firm? Is it primarily the customer who

Information flows at each stage of the customer journey

	Why does the customer engage in the interaction?		How does the customer go about identifying, ordering, and paying for the desired product?				What products and services are provided to the customer?	
	Latent need	Awareness of need	Search for options	Decide on options	Order and pay	Receive	Experience good/ service	Post-purchase experience
Description of information								
Trigger								
Frequency								
Richness								
Customer effort								
Action by								
Improvement ideas								

is making the inference on how to fulfill the need, or does the firm help in that decision?

Each question captures one of the five dimensions of the information flow discussed in chapter 4 on the recognize, request, and respond cycle. For now, you can leave the last row of the worksheet (improvement ideas) blank. You will come back to this row in step 6 of this workshop.

Step 4: Identify the Deeper Needs of the Customer

In chapter 5, we discussed the usefulness of the why-how ladder as a tool to discover your customers' deeper needs. Recall that each of the boxes in the why-how ladder corresponds to a specific way of expressing a customer need. Customer needs at the bottom of the ladder are more focused; they address *how* a particular need could be fulfilled. We climb up the ladder by asking *why*: Why is that need relevant in the first place? What would be good about fulfilling this customer need? This is what gets us to the deeper, more fundamental needs of your customer.

Use worksheet 6-4, the why-how ladder, based on the needs of your customers. There is no single right answer here. Answers at different rungs of the ladder correspond to alternative value propositions that your firm can provide to the customer. If you only focus on the *how*—that is, the lower parts of the ladder—you are at risk of being only in a transactional relationship with the customer. The moment an alternative solution becomes available to the customer, one that is associated with a higher willingness-to-pay or a lower price, your customer will grab it.

On the other hand, there also is a risk of focusing too much on the *why*. The more often you ask *why*, the fuzzier the value proposition becomes. You move from "provide me with an appointment" to "see a

The why-how ladder

In the eyes of the customer, the purpose of the relationship with our firm is to . . .

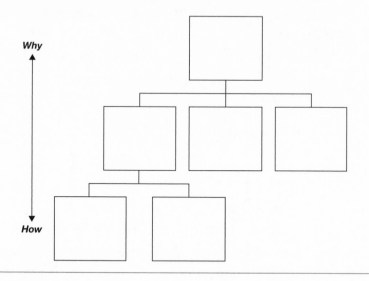

cardiologist" to "keep my heart healthy" to "have a longer and healthier life" to "be happy." We propose that the first steps of broadening the problem by asking *why* are helpful. But be reasonable, because too many *why*s and you'll find yourself searching for world peace and eternal bliss (and chances are you will not be able to move the needle on those objectives).

Step 5: Understand the Current Relationship with Your Customer across Separate (Repeated) Customer Experiences

As we discussed in chapter 5 (on the repeat dimension), there exists an important difference between a single episode of a customer experience and a connected relationship. In the worst case, your customer only has one experience with your firm and you will never meet again. Not much better is the scenario in which you provide multiple experiences to the same customer, but every time, you treat the customer as if you have never met before. In chapter 5, we laid out four different levels of customization that are enabled by the repeat dimensions. Let's work through these four levels for your organization.

In order to weave multiple customer experiences together into a connected relationship, the firm has to be able to identify the customer and retrieve information about the previous customer experiences. This is level 1 of customization. Answering the following questions will help you with this step:

- How do you identify the customer and connect him or her to prior customer experiences?

- Across which domains are you able to identify the customer? (E.g., stores or dealers in different geographies, online activity, or different agents from the same customer firm.)

- Is this identification requiring time and effort from the customer?

- Is this identification costly to your firm?

- What organizational incentives are in place (or what disincentives need to be removed) so that various parts of your organization share the information they have about a particular customer?

Beyond identifying the customer across multiple episodes of interaction, you want to improve the customer experience by learning more about this particular customer from one episode to the next. This is level 2 of customization. Ask yourself the following questions:

- How do we improve customization for a particular customer based on information that we have gathered about this customer?

- What feedback do we gather from the customer to understand whether a particular solution worked well?

- Can the customer make direct suggestions to us for how to improve our product or service?

- How do we share this information inside our organization?

As we discussed in chapter 5, learning not only happens at the level of an individual customer but also at the level of the population of customers that you face. Level 3 of customization relates to changing your available products and services to create more value. Ask yourself the following questions:

- How do we currently use population- (or market-segment-) level data to improve our product assortment?

- How do we currently use population- (or market-segment-) level data to refine features of existing products?

- How do we currently use population- (or market-segment-) level data to create entirely new products?

Lastly, use the insights you gathered from step 4 in this workshop, in which you worked on the why-how ladder, and ask how you can become a trusted partner of your customers. This is level 4 of customization.

- At what level in the why-how ladder are most of our transactions currently taking place?

- What would be alternative value propositions to the customer that are either more focused (*how*) or broader (*why*)?

Step 6: Identify New Opportunities to Reduce Customer Pain Points and Lower Fulfillment Costs

Based on the work that you did in the previous steps, the goal of this and the following step is to generate a list of opportunities to improve

(a) your current customer experiences and (b) what happens across experiences.

Such opportunities are a combination of the following elements:

- A customer pain point, a forgone revenue opportunity, or a waste of firm resources

- A change along any of the five dimensions of information flow at any of the steps in the customer experience or from one experience to another

Let us first focus on the willingness-to-pay drivers and pain points that you identified in step 2 and the information flows that you analyzed in step 3. To generate ideas, we find it helpful to go in both directions:

- Start with the willingness-to-pay drivers and pain points you identified and ask, "What could we do to relieve this pain point or to better address this willingness-to-pay driver?" And then ask yourself, "What information would we need that would allow us to implement our solution?"

- Or start with the information that you could potentially gather, and then ask yourself, "What pain point could we relieve with this additional information? Which willingness-to-pay drivers would we be able to address with this information?" To identify valuable information, it can also be helpful to ask yourself, "For what data would we be happy to spend a lot of money (and exactly how much?) because it would help us to increase our customers' willingness-to-pay or improve our efficiency?"

To generate ideas, it is sometimes also evocative to ask, using the different connected customer experiences we introduced in chapter 4:

- What could we do to create a better respond-to-desire customer experience?

- How can we build a better curated offering?

- For what aspects would our customers potentially value a coach behavior experience?

- Could we create an automatic execution experience for some of our services?

Recall that these different customer experiences affect different stages of the customer journey, as we illustrate in worksheet 6-5. Use this worksheet to keep track of your ideas and the required information. It is useful to draw arrows between the responses to pain points (or willingness-to-pay drivers) and the boxes containing the required information.

Once you have identified the required information for your various ideas, you can go back to your work of step 3 and fill in the bottom row of the corresponding worksheet: What changes do you have to make to your information-gathering activities in order to acquire the information you need?

At this point, we want you to come up with as many opportunities as you can.

An example for an opportunity that could come out of this step would look like this:

> Customers are often late when it comes to retirement planning. By tracking their spending patterns and their current assets and debts, our bank could engage earlier in the customer journey when the need is still latent, as opposed to waiting until the customer comes to us. A personalized video chat could be initiated by us and be used to engage the customer in this process by comparing her or his account with those of a peer group.

But it is not all about pain points and willingness-to-pay. The information flows in a connected relationship can also improve your efficiency, thus driving down fulfillment costs. This can happen in one of two ways. First, the information flow might already be happening, but it is happening manually. A dispatcher uses radio communication

Responses to pain points and required information

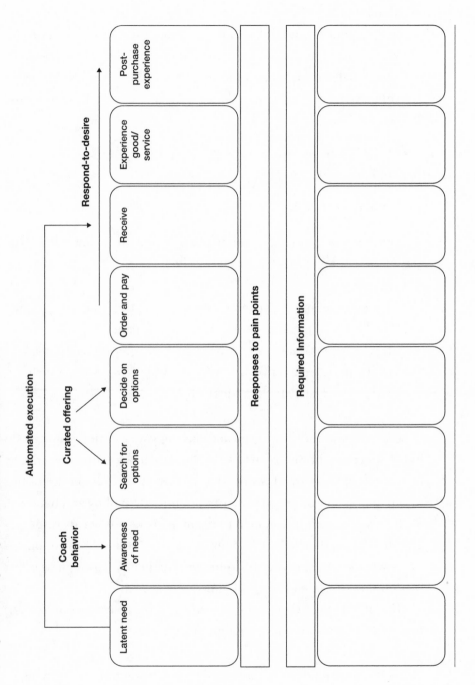

Automated execution

Coach behavior   Curated offering   Respond-to-desire

| Latent need | Awareness of need | Search for options | Decide on options | Order and pay | Receive | Experience good/ service | Post-purchase experience |

Responses to pain points

Required Information

to organize a passenger pickup; a Disney waiter gets a food order from a customer by handing out a menu, coming back, and then taking the order; or a patient goes through registration in the hospital before receiving care. As important as this information might be, it is definitely not adding value in the eyes of the customer and it unnecessarily drives up fulfillment costs. Specific questions to be asked include the following:

- Where is transactional information exchanged in person, via phone, or via fax?

- Where do you still use manual data entry?

Consider the following example of an opportunity that reflects the automation of an existing information flow:

> Patients coming to the clinic spend a long time checking in and dealing with paperwork. By tracking the patient's phone through geolocation and using facial recognition technology, check-in becomes effortless, which is more convenient for the patient and cost efficient for the clinic.

Beyond streamlining or even automating your existing information flows, there might be opportunities for efficiency enhancements by connecting previously unconnected parties. We saw examples of this in our discussion of grocery shopping and ride hailing in chapter 2. Such new connections can help to balance workloads across resources and thereby further reduce fulfillment costs. How might the opportunities you identified earlier be improved further by adding new information flows between previously unconnected parties?

How about connecting patients with care providers just in time, as opposed to only when the appointment was made?

> Often, patients don't show up for their appointments, leaving the provider or the equipment unused. Whether a patient will

show up on time, late, or not at all is only known at the last minute. By tracking the patient's phone through geolocation and using local traffic information, an expected arrival time can be calculated. If the patient is not within five miles of the hospital thirty minutes before the appointment, a reminder message is sent or the appointment slot is released.

Step 7: Find Ways to Utilize Information Gathered from Repeated Interactions to Improve the Recognize-Request-Respond Cycle

A key concept behind the repeat dimension of a connected relationship is that you learn from one customer experience to another. In step 5, you analyzed what you are currently doing for each of the four levels of customization that relate to the repeat dimension. In this step, we ask the natural follow-up question: How can you improve on each of these stages?

Questions you can ask yourself include the following:

- How can we improve our ability to keep track of an individual customer and to aggregate information across any touch point we have with this customer?

- How can we improve the internal incentives, or what changes to the organizational structure do we have to make in order for information sharing to happen?

- What have we learned about this customer and the customer's associated needs that would allow us to play a more active role in offering solutions to those needs? How could we use this information to better customize the product or service in the future?

- What have we learned from serving similar customers that would allow us to change the kinds of products and services that we will include in our lineup in the future?

- What have we learned about our customers that might give us insights that even our suppliers do not have about what future features or services are valuable to our customers?

- What efficiency improvements can be identified based on learning from the experiences of one customer and from the experiences of the population as a whole?

- Resulting from these adjustments, what deeper needs of our customers are we able to address (coming back to the deeper needs that you identified in step 4)?

On worksheet 6-6, write down a concrete example of a customer and a sequence of experiences that this customer has with you, then identify how you could use the information that you gather to improve customization, optimize your product and service offerings, create new products and services, improve efficiency, and eventually climb up the hierarchy of needs.

The output of this should take the form of an opportunity such as the following:

> Many students in MBA finance courses struggle when learning to compute net present values or discounted cash flows. By tracking the ability of a student to handle such questions across multiple homework assignments and potentially even multiple courses, a smart textbook can determine whether the student does not understand the definition of these concepts or whether he or she is struggling with more foundational content such as computing compound interest rates. If needed, the student is shown a tutorial for compound interest rate calculations.

Learning from repeated customer experiences

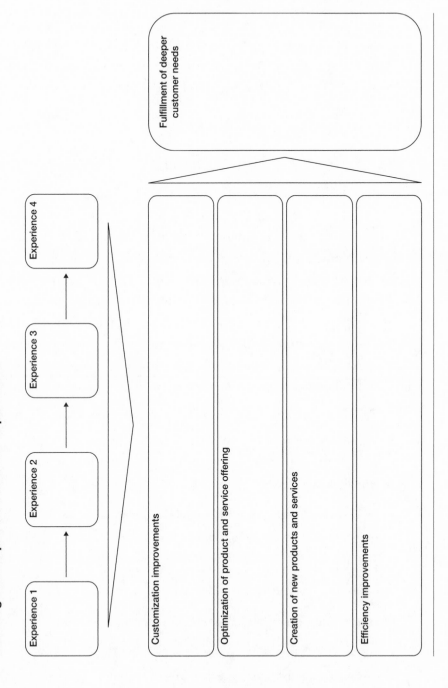

Experience 1 → Experience 2 → Experience 3 → Experience 4

Customization improvements

Optimization of product and service offering

Creation of new products and services

Efficiency improvements

Fulfillment of deeper customer needs

This opportunity might be further expanded to capture population-level learning:

> By tracking the homework submissions for an assignment, the teaching team can adjust the lecture plan to spend more time on content on which the general student population tests poorly.

Step 8: Assess Your Data-Protection Policies to Maintain Trust with Your Customers

To create a sustainable competitive advantage using a connected strategy, it is important for firms to create and maintain a trustworthy relationship with customers. Only if you safeguard your customers' data and use it in transparent ways will the customers continue to allow you to move up their hierarchy of needs. You need to make sure that you monetize your customer data in accordance with the relevant law, in agreement with your customer, and in a way that it is protected from criminal abuse. Some of the key questions you need to ask yourself include:

- What procedures do we have in place to stay informed about data protection and privacy regulations in all the areas in which we are active?

- How do we keep up with how public opinion is changing with respect to these issues?

- How do we currently obtain customer consent? How transparent is it to our customers what happens to their data?

- What do we do to keep the data current and accurate?

- What do we do to keep the data safe, and under what conditions do we notify customers of any breaches?

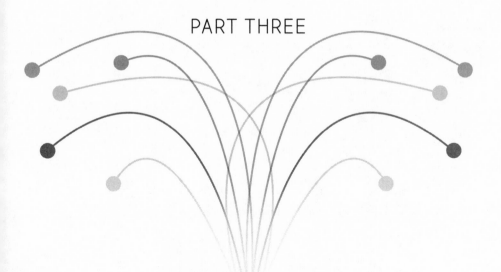

PART THREE

CREATING
CONNECTED
DELIVERY MODELS

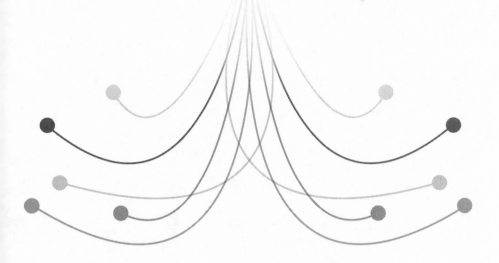

7

Designing Connection Architectures

For a successful connected strategy, a firm not only needs to create connected customer relationships but also has to create these relationships in a cost-effective way. To succeed, a firm needs a connected delivery model, which consist of three elements:

1. The connection architecture (equivalent to a series of pipes connecting the firm, its customers, and its suppliers)

2. A revenue model (how the firm can benefit from what flows through their pipes)

3. A technology infrastructure (how well these pipes are lubricated)

In this chapter, we focus on the connection architecture—a design aspect for which options have been proliferating.

Here's how the question of connecting the firm, suppliers, and customers can play out in the realm of finances. Consider Jane, a filmmaker seeking to create a documentary about polar bears. Today was an exciting day in Jane's life: she finally drafted an initial budget. To celebrate having created a budget, and to solicit some advice on how to find financing, she went out with her two best friends to a local bar. As usual, she wanted to split the tab at the end of the evening, but she realized that she accidentally left her wallet at home. Her friends paid for her, and she naturally promised to repay them. That interaction started the group talking about money and how they all need to save for retirement. As Jane walked home, many numbers were swirling in her head: the $25,000 funding for her film project, the $40 she owed her friends for the drinks, and the $300,000 she realized she might be short in retirement savings.

As this little vignette shows, customers have a wide variety of financial needs. Customers sometimes need to borrow money (to buy a house or create a new venture). Customers need to manage transactions (moving money from A to B, be it from the customer's bank account to a friend or to a restaurant), and customers need to save for retirement. Historically, customers relied on their bank to help with most financial needs. A customer opens a checking account and gets a credit card issued by the bank; the customer talks to a loan officer to get financing for projects; and the customer talks to a financial adviser of the bank, who recommends a mix of mutual funds for retirement, including funds managed by the bank itself. But the banking industry is currently going through a major transformation created by a movement known as fintech (financial technology).

Different connection architectures lie at the heart of some of the most profound disruptions.

Connected Producers

All connection architectures start with information flowing from customers to firms. These customers might be individuals or businesses. Thus every connection architecture works in both a business-to-consumer (B2C) setting and a business-to-business (B2B) setting.

Connection architectures differ in how the firm connects back to the customer. In the most straightforward connection architecture, it

FIGURE 7-1

Connection architecture for a connected producer

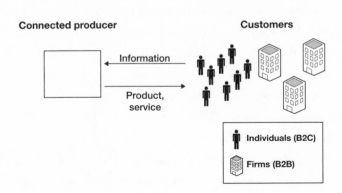

is the firm itself that produces the product or service that fulfills the customer demand. We call this type of business a *connected producer*. This is summarized in figure 7-1, which captures the key connections or pipes between organizational entities in the value chain.

In the financial services sector, for example, traditional banks that create connected customer relationships would be considered connected producers. The key connection is between customers and the bank, with the bank itself creating all the key services. Connected producers can focus on one connected customer experience or offer multiple experiences. For instance, a bank may try to make the loan application and approval process as fast as possible (respond-to-desire), make suggestions to the customer regarding how to invest based on the customer's income and future needs (curated offering), send the customer reminders to rebalance the portfolio (coach behavior), and automatically transfer funds between the customer's savings and checking accounts to avoid overdrafts (automatic execution).

Connected producers face competition from other connected producers. But some competitors of big banks do not look like big banks at all. Consider the loan process. Firms such as OnDeck and Earnest have focused on a respond-to-desire user experience. In addition to using traditional credit scores, they draw on over one hundred external data sources to determine creditworthiness, including social data, public records, and transactional data. This results in over two thousand data points per loan application and frequently allows the companies to provide a decision almost instantaneously, beating traditional bank application and approval processes, which are usually lengthy. In this space, OnDeck has focused on small business loans, while Earnest is a provider of personal loans, with a focus on refinancing student loans.

Notice, however, that at the thirty-thousand-foot level, they are operating exactly as big banks. The main connection is between the customer and the firm, in which information flows from the customer to the firm, and products and services flow back from the firm to the customer.

The architecture of connected producers appears old-fashioned. However, two recent connectivity trends are helping connected producers to innovate in how they serve their customers. First, today's connected producers increase the interaction frequency substantially, transforming big, episodic transactions into many smaller customer experiences. Second, the interaction's overall richness, in terms of information exchange and product customization, has intensified. For some examples of these developments, let's look beyond financial services.

Connected producers of transportation now provide user experiences far beyond selling cars. The world's largest car-sharing service is called car2go. It was created by a traditional producer, Daimler, and offers Daimler vehicles exclusively. Members have access to more than fourteen thousand vehicles in twenty-six cities in eight countries. Many of these vehicles are small, two-passenger Smart Fortwo vehicles (Smart is owned by Daimler). Users pick up cars wherever they are parked in a city, check them out using a smartphone app, do not have to refuel, return the car to any approved parking spot on the streets, and pay based on rental time.

Daimler's car2go is an example of a connected producer that is operating mainly in the respond-to-desire mode. Customers send their requests to the firm, and the firm responds immediately via an app that helps locate the nearest car that most closely fulfills the customer's needs. Better connectivity made this possible: GPS devices track a car's location; mobile apps tell customers where the cars are and eliminate the need for face-to-face payment. This makes using an expensive vehicle more efficient. What sounds like a cool application of the sharing economy corresponds to our definition of a connected producer architecture.

Competitors such as General Motors, Volkswagen, and BMW have followed suit and created their own car-sharing services. As we've noted before, just creating connected experiences will not be enough to create a long-term competitive advantage. Others will inevitably copy you, and the question remains whether you are able to drive a bigger wedge between customer willingness-to-pay and cost than your competitors. To achieve larger scale, BMW and Daimler decided to merge their mobility platforms in 2018 to create one car-sharing service.

Similar to the financial services industry, the insurance industry is seeing firms that are creating connected strategies. While there are many new entrants into what has been dubbed insurtech (insurance technology), some of the existing connected producers have been very innovative as well. Consider Progressive, one of the largest American automobile insurance companies. Buying auto insurance has historically been a very episodic customer experience. A driver calls the insurance or goes to an agent; an underwriter analyzes the driver's risk profile based on age, zip code, vehicle type, and past accidents; and the company then provides a quote for an insurance premium. In contrast, Progressive customers can agree to use an app or plug in a small monitoring device in their vehicle. The app or device monitors the length and time of day of travel, as well as the driving behavior including hard braking, fast acceleration, or aggressive lane changes. Once the data-collection period is complete, Progressive uses the data to deliver a tailored car insurance quote to the driver. But the customer experience doesn't end there. Progressive is also creating a coach behavior experience by providing customers with feedback about their own (or their children's) driving behavior. This allows customers to improve their driving, which could further lower their rates. Moreover, this allows Progressive to climb up the hierarchy of needs of its customers: it addresses the deeper need that drivers really have for safety and security, rather than for having car insurance.

Many software companies, such as Salesforce and IBM, also fall into the category of connected producer. Their tighter connection with their clients allows them to customize their offerings and anticipate future needs (e.g., with respect to cybersecurity). Likewise, Google, with its search engine, Gmail, and Google Photos, is a connected producer. The fact that Google does not charge for these products does not detract from its status as a connected producer who has established its own ecosystem of services. (We will return to the revenue model of Google and similar firms in chapter 8.)

Finally, consider the example of Disney's MagicBand, which we discussed in chapter 1. The band incorporates respond-to-desire elements by enabling easy access to hotel rooms and rides and easy payment for merchandise and food. Through its app, Disney incorporates elements of curated offering with suggestions, and it engages in coach behavior by redirecting visitors to less crowded rides. Not surprisingly, other entertainment firms are developing similar technologies. For instance, Carnival, the largest cruise line operator in the world, has developed a device called an Ocean Medallion that can be carried in a pocket or worn as a pin. This medallion, just the size of a quarter, automatically opens the door of the passenger's stateroom (no need to tap a sensor), makes it faster and easier to embark and disembark, and facilitates payment for items purchased on board. Linked to an app, it can help family members find one another on expansive ships. The app can be used to order food for delivery wherever the passenger plans to eat. Servers know who ordered the food because the passenger's photograph appears on their handheld device when they get close to the passenger's medallion. Passengers can upload their preferences before boarding, which enables Carnival to offer tailored activities. Carnival and Disney are both using a connected producer architecture. The connection is between the customer and the firm; it just has become a very high-bandwidth connection.

Connected Retailers

Connected producers face competition not only from other connected producers but also from firms implementing different connection architectures. For instance, in the realm of investment advice, companies like Wealthfront and Betterment mainly use algorithms to assemble a customized portfolio for each investor that comprises exchange-traded funds or low-cost mutual funds managed by others. We call these firms *connected retailers*—firms whose main role is to showcase, curate, and deliver products from suppliers to customers. Connected retailers receive information from customers and then create a connection between suppliers and customers. This connection runs through the connected retailer, as the firm is actively involved in moving the product from the supplier to the customer (figure 7-2).

Like connected producers, connected retailers can focus on one connected customer experience or on a range. For instance, Wealthfront

FIGURE 7-2

Connection architecture for a connected retailer

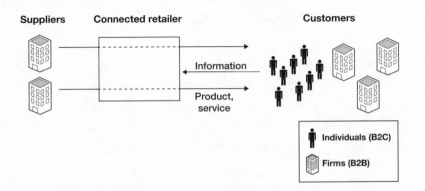

and Betterment make creating a customized investment portfolio quick and easy by doing everything online (respond-to-desire); they provide a curated offering by allowing customers to state various investment goals and then customizing a portfolio that aggregates over those goals; and they have an automatic execution experience through services like automated loss harvesting.

Amazon's practice of shipping out of its own warehouses is perhaps the most well-known example of a connected retailer. Customers interact with the company mainly through a respond-to-desire connection. However, Amazon has clearly increased curated offerings through "customers like you purchased" recommendations and automated offerings, such as subscription programs where customers periodically receive products like toothpaste or detergent.

Many connected retailers create customer value through a curated offering customer experience. While the internet has given customers access to practically every product and service available worldwide, such choice is overwhelming and requires a lot of customer time to make good decisions. Therefore, curation can create a lot of value. Connected retailers offering subscriptions to curated boxes of merchandise have sprung up across hundreds of product categories, from food to cosmetics to pet supplies. Just a sampling of subscription services starting with the letter *B* is illustrative. From Bad Ass Mom Box (jewelry and beauty products) to BarkBox (dog toys and treats) to Busy Bee Stationery (stationery), all of them have the same user experience and the same connection architecture. Firms purchase items from their suppliers and assemble them as a curated offering tailored to each customer.

We have already mentioned curated meal subscription box retailers such as Blue Apron and HelloFresh. Several other players have emerged in this space as well, trying to differentiate themselves by focusing on particular cuisines or preparation time. Competitors include Purple Carrot (vegan), PeachDish (Southern cuisine), Sun Basket (paleo and gluten-free), Green Chef (certified organic), and Gobble (preparation time of ten minutes using only one pan).

Like some connected producers who coordinate customers to utilize resources more efficiently, several connected retailers have a similar business model. Car-sharing services like Zipcar (now owned by the Avis Budget Group) fall into this category. Unlike car2go, which is operated by the firm that manufactures the shared cars, Zipcar buys or leases cars from various manufacturers, providing broader choice. (For another example from the mobility industry, see the sidebar.)

Rent the Runway has a similar business model for a very different product category: designer dresses. Because these dresses are expensive and often worn only once, they are a great example of underutilized resources. With Rent the Runway, women can rent dresses for four or eight days at a price much lower than that of purchasing them. Each rental arrives in two sizes, ensuring the dress will fit. Customers return them via UPS, and Rent the Runway handles the dry cleaning.

Both Rent the Runway and Zipcar are exploiting an important customer trend. What many customers actually desire is access to a product or service, not necessarily ownership. Especially when needs vary over time (today I'd like to drive a convertible; yesterday, I needed a pickup truck; and tomorrow I need a minivan) and the assets are very expensive, it has been traditionally very difficult for a customer to match needs with solutions (and the outcome is usually a compromise like a family sedan). Connected retailers who rent, rather than sell, products allow customers to achieve a better match between their needs and solutions at an affordable price. One might refer to these cases also as instances of the sharing economy, but from the perspective of a firm that has to build a connected delivery model, we find it more useful to classify them as connected retailers.

It is interesting to note that some firms that started out being connected retailers, such as Netflix, Amazon, and Zalando, have used their customer data to enter into the production of movies and private-label items and have become connected producers as well. Their direct connection to the end customer provided these firms with deeper knowledge about customers than any of their suppliers possessed, allowing the firms to successfully backward integrate. In sum, to

E-SCOOTER AND BATTERY SHARING

Mobility in cities has been changed not only by ride-hailing services but also by a range of bike and scooter rental options (both electric kick scooters and motorized scooters). Consider Coup, an electric (motorized) scooter rental company operating in Berlin and Paris, and owned by Robert Bosch, a German engineering and electronics company that is one of the world's largest manufacturers of automobile components. Customers can rent an e-scooter (including a helmet) using Coup's app. The scooters are GPS equipped, so they can be picked up and left anywhere within the city limits. Payment is automatic and based on usage time. One common issue with such sharing services is that scooters end up in places where they aren't needed, such as in front of bars on Sunday mornings. Coup has crowdsourced the solution to this problem. It offers free riding time to volunteers who are willing to move the inventory of bikes to desired locations. Interestingly, the e-scooters used by Coup do not have any Bosch parts in them. In order to move faster in this space, Bosch decided to use a scooter produced by Gogoro, a Taiwanese manufacturer, in essence making itself into a connected retailer like Zipcar. Gogoro, in turn, pursues a connected producer strategy in Taiwan. It sells, rather than rents, its scooters there with a monthly subscription fee that allows riders to change batteries at swapping stations located throughout Taipei. Riders can find the stations and reserve their batteries through an app. Gogoro believes customers do not want to share their scooters, but they are willing to share their batteries. When charging its batteries at the stations, Gogoro can take advantage of low electricity prices at times of low electricity demand.

create a connected strategy, firms can employ more than one connection architecture.

Connected Market Makers

While Wealthfront and Betterment are actively involved in assembling and managing their clients' portfolios, other firms have roles primarily in establishing a connection directly between a customer and a supplier. For instance, LendingTree is an online lending exchange that connects customers with multiple lenders that compete for business, allowing the customer to select the best supplier for his or her needs. We call these firms *connected market makers*—firms that create a direct link between supplier firms and customers but that aren't involved in handling the product or service. Connected market makers control the connection from the customer to the supplier, but they do not own what flows back from suppliers to customers. Connected market makers thus rely on existing suppliers to fulfill the needs of customers.

This is illustrated in figure 7-3. Note that unlike the connected retailer, who gets directly involved in handling the product, a connected market maker establishes the connection and potentially vets the suppliers but is otherwise hands-off in how the supplier fulfills the need, thereby carrying neither inventory nor financial risk. Also, the number of suppliers that can be connected to customers can be very large, since the connected market maker is not responsible for holding inventory or directly managing the stream of goods and services from the suppliers to the customers.

Wallaby Financial is another example of a connected market maker in financial services. If a customer stores all of her credit and loyalty cards on its platform, Wallaby's app will recommend the optimal card for each purchase based on the nature of the purchase and the attributes of the cards in her mobile wallet. Thus, the company can manage the customer's cards to minimize fees paid while maximizing rewards and discounts. Using this data, Wallaby recommends new credit

FIGURE 7-3

Connection architecture for a connected market maker

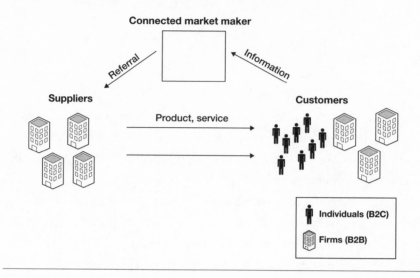

cards for more savings. In this case, Wallaby creates new connections between credit card issuers and customers, earning a referral commission.

The distinction between connected retailer and connected market maker is subtle but important. By focusing entirely on connecting businesses and customers (without purchasing products or capacity), connected market makers are less capital intensive. Consider Amazon again. As a connected retailer, it invests in warehouses and inventory. But with products that are sold through Amazon Marketplace, the company acts as a connected market maker by taking the customer order and redirecting it to the seller, who then takes care of the order.

Expedia and Priceline work similarly. They neither own airplanes nor commit to purchasing seats on flights or rooms in hotels. They simply find customers looking for air travel or accommodations and then connect them to the right airline or hotel. Connected market makers exist both in the B2C world and in the B2B world. For instance, Iron-Planet is a connected market maker for used construction, transportation, and agricultural equipment, a market estimated to be worth around $300 billion. IronPlanet connects industrial buyers to industrial sellers of heavy equipment. (For another example of a connected market maker, see the sidebar.)

So, if the market maker is entirely hands-off in the transaction, what value does it create? Wouldn't it be better if the customer connected with the supplier directly? Market makers can fulfill two functions:

> They can provide curated offerings and, free from the need to make any investments in fulfillment, can do so on a bigger scale. For example, OpenTable can connect patrons looking for restaurant tables with pretty much any restaurant in the country, a level of choice and access that is impossible (or really, really inconvenient) for a customer to replicate.

> Market makers can ensure that the products and services on their platform are top quality and provided by reputable sellers. While the individual customer interacts with a business only once, the market maker does so repeatedly. The vetting can be done by the market maker itself, as in the case of Sweeten, a market maker that connects reputable contractors to customers with significant renovation projects, or through reputation scoring done by previous customers, as in the case of Angie's List, a crowdsourced directory that has accumulated millions of reviews for all kinds of local businesses and that connects those businesses to customers.

Given that initial entry costs for a connected market maker are relatively low, we have seen many start-ups using this connection architecture. The downside of low entry costs, of course, is that entry is easy for everyone. As a result, we have seen rampant imitation, leading to ever growing costs of attracting participants on both sides of the market. Market makers have to provide increasingly better incentives to suppliers to convince them to join, while at the same time customer acquisition costs are rapidly increasing. Google and Facebook

THE PANDORA OF THE ART WORLD

The global market for art is more than $60 billion annually, yet it is very fragmented. Art is sold in auction houses and in thousands of art galleries around the world. This makes the market for art in the range of $50 to $100,000 very local and inefficient: art collectors buy what they find in galleries in their hometown but are usually unaware of what is available in other galleries. Artsy, a connected market maker, is changing this by connecting galleries and auction houses to art collectors around the world. More than eight hundred thousand pieces of art from more than two thousand galleries representing eighty thousand artists can now be found on Artsy. Such a huge catalog of art can be overwhelming. Artsy adds value not only by providing a marketplace but also by creating a curated offering customer experience. Similar to what Pandora did with music and Netflix with movies, Artsy has created a classification system (the Art Genome Project, inspired by Pandora's Music Genome Project) that describes each piece of art along more than one thousand dimensions that Artsy calls "genes." The dimensions include characteristics such as historical movements (e.g., contemporary Turkish art, pop art), subject matter (e.g., food, shadows), visual qualities (e.g., asymmetrical, blurred), medium and techniques (e.g., animation, collage), and materials (e.g., aluminum, gemstones). This classification system allows potential buyers to explore the space of available art in highly nonlinear yet related ways. For instance, a buyer who likes Andy Warhol, whose work, however, is outside the buyer's budget range, could be guided to younger artists whose work shares some dimensions with Warhol's and might be photographs rather than paintings.

have become the de facto new "landlords." Market makers don't pay the traditional rent for stores or warehouses, but they pay Google and Facebook and others for customer leads. Thus, as some industry observers have noted, customer acquisition costs have become the new rent.

Crowd Orchestrators

What are alternative sources of funding for a project like filming a documentary? For a loan, one option is Prosper. Borrowers request loans between $2,000 and $35,000, and individual investors invest as little as $25 in loans they select. Prosper handles the loan servicing on behalf of the matched borrowers and investors.

Whereas connected market makers create a connection between customers and supplier firms (e.g., banks), Prosper connects customers to individuals who serve as suppliers (of funds in this case). Not only does Prosper rely on individuals to serve as suppliers, but Prosper was essential in creating this funding source—a crowd of individuals—in the first place. Prosper thus orchestrated a crowd: it created a new set of connections among previously unconnected individuals. Consequently, we call this connection architecture *crowd orchestrator.* Just as with connected market makers, here the firm focuses on creating connections instead of producing or handling products directly. This time, however, the connections are made between individuals and customers, not between existing firms and customers (figure 7-4).

Kickstarter is another crowd orchestrator and fits the profile in figure 7-4. It connects individuals who want to support projects or want to pre-buy products that are still in the creation phase (like the documentary on polar bears). Though each individual contribution might be small, Kickstarter has collected in its first nine years over $4 billion in funds and supported over 154,000 projects, all of which happened without the involvement of a bank or venture capitalist. Kickstarter

FIGURE 7-4

Connection architecture for a crowd orchestrator

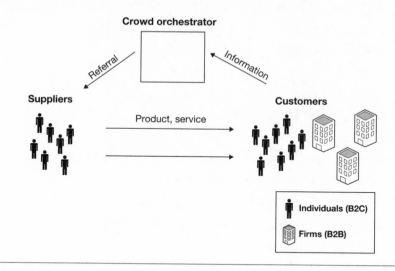

not only funds individual projects but is also quite often used to fund startups. Banks and other financial institutions now face competition from individuals banding together on these platforms. Where once banks might have counted on people financing creative projects with personal or high-interest credit card loans, some of this market has been taken by crowdfunding.

While both connected market makers and crowd orchestrators are similar in that they directly connect customers to suppliers, there are key differences between these connection architectures. A customer who transacts via a connected market maker would quite likely have made a similar, although possibly less optimal or convenient, transaction without the market maker: the customer still would have made a reservation at a restaurant (without OpenTable), booked a hotel room (without Priceline), or a bought a plane ticket (without Expedia). These transactions would have happened because the suppliers that connected market makers connect already existed. In contrast, a crowd orchestrator creates new supply by allowing individuals who otherwise would not have participated as suppliers to enter the marketplace. By creating connections between these individuals and customers, a crowd orchestrator is essentially creating a new market. Without Prosper, it is very unlikely that a borrower could find individuals who would give him twenty-five dollars each; without Airbnb, it is very unlikely that a traveler would have found an apartment to rent for a single night. (For other examples of crowd orchestrators, see the sidebar.)

One particular design consideration for crowd orchestrators is how much control to give to both customers and suppliers. Can customers choose individual providers? Can suppliers set their own prices? Overall, the more control a crowd orchestrator exerts, the more the customer experiences the work performed by the crowd as one coherent virtual firm. It is not the individual provider of the product or service that matters when using Lyft, Uber, Instacart, or others. In fact, the customer has no way of choosing who will fulfill the request, and it is the crowd orchestrator who sets the prices, not the individual suppliers.

In contrast, with less control, there is more variety in the market. You might not value that variety when it comes to traveling from A to B—for a short ride, a car is a car as long as it is clean and basically safe. The same is not true for a vacation home. In such settings, variety of choice is valued by the customer, and so imposing control ("Your doors have to be white and you need to charge one hundred dollars per night") would be useful neither to the customer nor to the provider of the home.

As with other connection architectures, we see that over time some firms operate with more than one architecture:

> Airbnb started out as a platform where travelers could find a place to stay while traveling by engaging in a short-term rental agreement with the owner. The owner acquired the real estate with the primary motivation to use it for herself and only hosted Airbnb guests to defray costs. That makes Airbnb a crowd orchestrator. But increasingly, rentals listed on Airbnb are owned by commercial real estate companies that rent their properties 365 days a year, using Airbnb as a sales channel. For

NOT-FOR-PROFIT CROWD ORCHESTRATORS

Connected strategies can be employed successfully not only by profit-seeking firms but also by organizations that have other objectives. The connection architecture of crowd orchestrator can be a powerful tool to connect individuals in need with individuals who want to help. Consider DonorsChoose.org. Public school funding in the United States has been very tight for many years, creating deficits in almost everything from art supplies to books and lab equipment. Teachers with needs in their classroom can submit a request to DonorsChoose.org. Its staff will vet and post the projects, including a detailed financial overview of how funds are being spent. Individual donors, and some corporate partners, then pick projects and contribute funds. Once a project is fully funded, the staff at DonorsChoose.org purchases the requested items and ships them directly to the school. In turn, each donor gets a thank-you letter from the teacher, photos from the classroom, and a report of how the funds were spent. Since 2000, DonorsChoose.org has connected more than 3.5 million donors to fund more than 1.25 million classroom projects, totaling more than $766 million in funds raised. Without this crowd orchestrator, it is hard to see how these donors and teachers would have connected with each other.

those customers, Airbnb is also becoming a connected market maker.

The "mother of all platform models" is the auction site eBay. Again, when an individual sells his old lawnmower on eBay, the company is acting as a crowd orchestrator. But when the local hardware store uses eBay as a distribution channel, eBay is a connected market maker. A similar development can be seen at Etsy, which started out as a marketplace where individuals could sell handmade items and now also offers factory-made items.

Crisis Text Line is another crowd orchestrator in the not-for-profit sphere. Crisis Text Line connects people in personal crises to trained volunteer crisis counselors who can work remotely anywhere with a computer and an internet connection. All conversations on the crisis-intervention hotline, lasting on average one hour, are conducted via text messages, the preferred medium for many teenagers. Crisis Text Line provides training to volunteers and has invested heavily in technology that allows texts to be automatically reviewed for severity, so that imminent-risk texters are responded to first. Volunteer crisis counselors are supported by full-time paid staffers who have advanced degrees in mental health or related fields. Since its launch in August 2013, Crisis Text Line has exchanged more than eighty-six million texts with people in need. While it started in the United States, it has replicated its system now in Canada and the United Kingdom. Given its extensive data, which links crisis topics, time, and geography (e.g., depression peaks at eight o'clock at night, anxiety at eleven o'clock at night, self-harm at four o'clock in the morning, and substance abuse at five o'clock in the morning), it provides access to its anonymized and aggregated data for free to police departments, school boards, policy makers, hospitals, families, journalists, and academics. Again, without this crowd orchestrator, the connections between people in need and people with good listening skills would not have been made.

Peer-to-Peer Network Creators

The other problem Jane had to solve in our little vignette was choosing to pay the bartender for the drinks or to reimburse her friends for the money. Even without a wallet, she could have settled accounts through her smartphone using a service like Venmo. Once a customer has signed up with Venmo, payments can be made by individuals with only a cell-phone number or an email address, enabling transfers regardless of an affiliation with a bank and the existing payments infrastructure.

Venmo is a *peer-to-peer (P2P) network creator.* In contrast to crowd or-chestrators, where it is clear that one individual is the supplier and the other is the customer, with P2P network creators, individuals might change roles frequently, as in the Venmo example. Today, we send money via Venmo to you. Next month, the transaction might go in re-verse. We are simply part of the same payment network. Venmo, owned by PayPal, now handles billions of dollars.

Yet another P2P network creator in the financial services space is TransferWise, which focuses on international money transfers. Send-ing money across country borders remains an expensive undertaking. TransferWise realized that if Mario, who lives in country A, wants to send one hundred dollars to Madhav, who lives in country B, and Jamini, in country B, wants to send the equivalent of one hundred dol-lars to Juanita, in country A, that result can be achieved by transfer-ring one hundred dollars from Mario to Juanita, who both live in country A, while transferring the equivalent of one hundred dollars in local currency from Jamini to Madhav, who both live in country B. That results in two domestic transfers, which are much cheaper to ex-ecute than two international transfers. By creating new connections between the various parties, TransferWise substantially reduces costs.

P2P network creators are remarkable organizations, often connect-ing millions of individuals. Moreover, they can present a threat to ex-isting businesses. Banks used to love domestic and international money transfers because they carried hefty fees. Now, these important reve-nue streams are being significantly affected because these competitors are utilizing completely different connection architectures.

Foreshadowing the next chapter, on revenue models, we find it help-ful to think about three different types of P2P network creators based on how they monetize their connection architecture:

> ***Transaction or membership revenues:*** While not as expensive
> as traditional banks, TransferWise does charge a transaction
> fee. Venmo, though it does not charge a fee for transactions in-
> side the network, does make interest income on the capital it

circulates and charges customers when they use credit cards to make payments. Membership revenue is the source of income for dating portals such as Match.com. Lonely hearts pay Match.com a fee to be connected to each other, which sometimes results in happy couples but is always cash in the bank for the company.

Fees for access to the information that is created in the network: Wherever there is content and traffic, there is income-generating potential, often from advertising. YouTube started out as a P2P platform for sharing videos. It now makes a fortune with commercials, selling access to finely calibrated audiences on the network (e.g., targeting specific ads to viewers of certain programs). LinkedIn allows people to join the network for free but sells access to the information to potential employers.

Revenues from complementary products: In addition to acting as a connected producer of running shoes, allowing customers to upload and analyze their running data, Nike acts as a P2P network creator by supporting virtual running clubs. These free clubs create a community of runners who encourage each other to run more. That's good news for a company selling running shoes.

We illustrate the P2P network creator architecture in figure 7-5. Individuals are connected to each other via the network, with most participants serving as both senders and receivers of whatever the network is designed to facilitate (money, information, etc.) For some network creators, revenues are generated "within the box" of the network. For other network creators, revenues are generated by selling access to information that is created within the network. Finally, some network creators are able to use the network they create to drive up the willingness-to-pay that customers have for other products and services they offer—that is, they use the network as a complementor.

FIGURE 7-5

Connection architecture for a P2P network creator

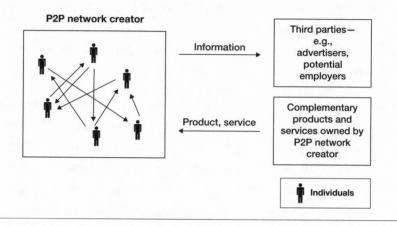

The Connected Strategy Matrix

In the second part of this book, you learned about creating different connected customer experiences. We introduced four of them: respond-to-desire, curated offering, coach behavior, and automatic execution.

In this chapter, we introduced five different connection architectures: connected producer, connected retailer, connected market maker, crowd orchestrator, and peer-to-peer network creator.

Each connection architecture can be used to create different customer experiences. With four connected customer experiences and five connection architectures, we can create a matrix that has the customer experiences on one axis (the rows) and the connection architectures on the other axis (the columns) (figure 7-6). We call the resulting matrix the *connected strategy matrix*.

The purpose of the connected strategy matrix is twofold. First, it can serve as a framework to help you understand both your own activities and those of your competitors. Where are your competitors operating in this matrix? Where are new startups popping up? Because

FIGURE 7-6

Connected strategy matrix

	Connected producer	Connected retailer	Connected market maker	Crowd orchestrator	P2P network creator
Respond-to-desire					
Curated offering					
Coach behavior					
Automatic execution					

firms can create more than one customer experience and can operate with more than one connection architecture, they can play in multiple cells of the connected strategy matrix. The second use of the connected strategy matrix is as an innovation tool. By going through each cell and asking yourself, "If our firm had a strategy in this cell, what would it look like?" you have a very structured way to guide your innovation process. We will guide you through this process in more detail in the workshop in chapter 10.

Beyond Platforms: The Five Connection Architectures

Over the last few years, many executives we have talked to have expressed concern about being disrupted, not by their competitors but instead by companies that operate very differently from them. The verb *to uberize* is making it into dictionaries, and uberization is praised for its utilization of digital technology to dramatically increase the efficiency of an economic system by leveraging platforms and P2P interactions. Beyond Uber and its competitors, this idea is most commonly illustrated with the cases of eBay, Airbnb, Zipcar, Facebook, and many other companies we discuss in this book.

The threat of digital disruption is real, as anybody owning a cab or running a hotel will testify. But before getting too excited about the sharing economy and platforms (see the sidebar in chapter 1), we find it helpful to look at these phenomena a bit more carefully. In our view, details matter, so while Uber, Airbnb, and others all make for great business school case studies, we should not overlook the fact that they operate very differently from each other.

In this chapter, we discussed five connection architectures:

> *Connected producers:* You don't have to be a radical new startup or create a two-sided marketplace to have a connected strategy. Traditional producers such as Disney, Nike, and

Daimler have created connected strategies within parts of their businesses by changing how they connect to their customers and by moving from episodic interactions to continuous connected customer relationships.

Connected retailers: Traditional retailers ask customers to come to their store and to buy what they have. Connected retailers make the choice process and the ordering of, payment for, and receipt of the product much more convenient for a customer. From Amazon for books, and Netflix for movies, to meal-kit providers for groceries, connected retailers create a much closer relationship with customers, which allows customization and the reduction of pain points along the entire customer journey.

Connected market makers: These firms create a market by connecting supplier firms with customers. Examples include Expedia, Priceline, and Amazon Marketplace. Connected market makers are the bazaar operators of the twenty-first century. They neither buy nor sell; they just make sure the right buyer is connected to the right supplier. This approach sounds like the dream of every operations manager: it seems as if this approach requires almost no capital (capacity, inventory) while also being free of any operational risk. However, to succeed with this connection architecture, a firm needs to be able to attract both buyers and sellers (create a two-sided market) and provide them with liquidity and trust.

Crowd orchestrators: In contrast to connected market makers, the firms can't rely on existing suppliers. A key task of a crowd orchestrator is to mobilize individuals to serve as suppliers—for example, of driving services (Uber), shopping help (Instacart), accommodations (Airbnb), or financial resources (Kickstarter). The key challenge is to attract customers while the set of suppliers is still small. Once a critical mass is reached, though, two-sided network effects kick in: the more suppliers that are available, the more customers will come; the more customers that come, the higher the incentives for more suppliers to join.

Peer-to-peer network creators: These firms form and organize communities of users, blurring the lines of consumers and providers. Again, network effects play an important role in the sustainability of firms with this connection architecture. If the value a customer derives from a network increases with the number of participants (e.g., as the number of posted reviews increases), then larger networks will tend to attract more new users, increasing network size even more.

The connected strategy matrix that we introduced in this chapter helps you think through these differences. Moreover, it allows you to integrate your analysis of the connection architectures with the four connected customer relationships we introduced in the second part of the book. In the upcoming workshop in chapter 10, you will put this tool to work.

A final thought (for now) on the growth of platforms and networks. The connection architectures to the right side of the connected strategy matrix are enabled through advances in technology—ride sharing just does not work without mobile computing and geolocation. With technology further advancing, there is no reason to believe that this growth will stop any time soon. But that by no means implies that connected producers and connected retailers are the dinosaurs of connected strategy. For example, our opening vignette about Disney is one of a connected producer that focuses more on the rows of the connected strategy matrix (new connected customer experiences) than the columns. There is no "one size fits all" for connected strategies.

8

Revenue Models for Connected Strategies

The average American spends $384 per year on dental care between out-of-pocket expenses and insurance copays. That's about $10,000 on dental care over three decades, on top of the hassle of scheduling appointments, getting to them, waiting, and the pain of the care—a pain point if there ever was one.

Now, imagine that you are the CEO of an oral care or medical device company. You invent an amazing toothbrush that detects plaque and cavities before dentists or patients do. Using the insights from this book, your company makes it smart and connected. It guides the patient in the brushing process and schedules a dental appointment if needed. Your toothbrush, let's call it the Smart Connect XL3000, keeps customers' teeth so much cleaner that it cuts dental care costs in half and reduces time spent in appointments. Let's assume it costs $300 to produce and lasts five years, as long as the toothbrush head is replaced every six weeks. At what price would you sell the Smart Connect XL3000?

Before pondering your pricing proposal, be it $500 or $5,000, be it with a 50 percent gross margin or 20 percent more than the competition, note that the issue is about more than price. As far as our connected toothbrush, or any connected strategy, is concerned, the big-picture task is coming up with a revenue model.

We define a *revenue model* as one or several mechanisms that will compensate a firm by capturing some of the value that its products or services generate. In the case of our Smart Connect XL3000, customers benefit from healthier teeth, increased convenience from fewer appointments, and long-term savings on out-of-pocket costs and co-pays. If you sell the toothbrush for $400, almost all this value stays with the customer. Customers might love you for that, but you will only make a small profit that may not recover all the R&D investment you made to create the XL3000. If you price it at $5,000, however, very few will buy, destroying a lot of potential value.

Consider the following alternatives to selling the toothbrush with a fixed price:

> You could sell the toothbrush for $300 and then make your money by selling the replacement heads with a healthy profit margin, a model familiar to anyone using Gillette razors.
>
> You could offer a subscription model: for $10 a month, a new brush head is automatically shipped to the customer. In men's facial care, this is the revenue model that led to the emergence of the Dollar Shave Club, a startup later purchased by Unilever for $1 billion.

Both revenue models might be innovative compared to just selling at a fixed price, but they have nothing to do with connectivity. As can be seen by the examples of Gillette and the Dollar Shave Club, both strategies are used by the (poorly connected) existing razor companies.

As a firm with a connected strategy, you have a long-term relationship with the customer, including a high-bandwidth information exchange. What revenue models can you design that could not be replicated by a company relying solely on episodic interactions? Consider the following options:

You could charge the customer ten cents per minute of brushing time. Because of the connectivity, your firm measures this so you can use this information as part of your revenue model. (That creates a financial disincentive to brush, an issue that can be handled by making your guarantee contingent on certain minimal use.)

You could launch an optional app that helps the customer in her brushing behavior for a one-time fee of ten dollars or a monthly subscription fee. Such a coach behavior connected experience might alert the customer when the toothbrush has not been used for twelve hours, or when the right-handed customer brushes too much on the left side of the mouth and too little on the right side, or when the customer exerts too much pressure (note that such apps already exist for Bluetooth-enabled electric toothbrushes from Oral-B, for instance).

You could have a sensor at the brush head that automatically detects its deterioration and reorders a new head as needed, resembling the printer toner example from chapter 4.

These revenue models allow you to appropriate some of the extra value that your product creates for the customer (beyond the value that the customer is currently getting from using normal toothbrushes). Are there more alternatives? So far, we've only looked at dividing the value between your company and the customers. But what about other parties? Let's continue brainstorming other forms of revenue models:

Some dentists might not be happy about the Smart Connect XL3000, but others might pay you a referral fee if your toothbrush makes an appointment at their practices. How about insurers? You could give the toothbrush to the insurance companies for free and then ask to be paid a percentage of the savings relative to past patient expenses.

You could also collect data about brushing behavior, including what time your customer gets up in the morning and when (or what) he eats. You could sell this data to Starbucks (which could provide coffee just when customers are in their wake-up routine) or to the customer's life insurance company, alerting it that its customer seems to be on an unhealthy diet or is smoking, creating a strong incentive for the customer not to engage in these activities.

Finally, through your connection to the customers' bathrooms (not to mention their mouths), you could become a trusted partner in oral hygiene and have the Smart Connect XL3000 be the platform on which other oral care transactions are organized, earning referral fees when customers make purchases for toothpaste or dental floss.

The breadth of these models is already spreading across various industries, as the following real examples illustrate.

In the medication adherence domain, PillsyCap has developed a forty-nine-dollar pill bottle that reminds patients to take their medications or supplements. The bottle has a simple sensor to detect when it is opened and is connected to a cloud-based server. At AdhereTech, another startup with a similar technology, the pill bottles are given to the patient for free. Pharmaceutical companies and pharmacies benefit from the technology by selling more pills, and hospitals benefit from reduced readmission. Thus, a price of zero to the patient maximizes adoption and increases the value that can be shared among insurers, pharmacies, drug companies, and health care systems. But how can we be sure that the pill has been taken after the bottle is opened? Smart pill bottles have no answer to this question. The schizophrenia drug Abilify, a pill with an embedded sensor that tracks whether the medication has been ingested, has solved this problem. The pill's sensor is connected to a wearable patch that feeds data to a mobile application.

Similarly, consider Fitbit. Fitbit has emerged as a powerful brand for its wearable devices. Given millions of Fitbit users, the company has access to remarkable data. For example, it has access to 105 billion hours of heart rate data, six billion nights of sleep, and two hundred billion minutes of exercise. Even though all this data is depersonalized, it is still of enormous value. Fitbit is preparing to launch digital health tools for the detection of atrial fibrillation, sleep apnea, and other conditions.

The purpose of this chapter is to discuss revenue models for connected products and services. As you likely have noticed, we are using the health care space as a case study. Entire books have been written about revenue models, so our focus is on the unique opportunities for connected strategies. We do this in four steps:

1. We first provide a brief overview of revenue models and point to some of the key limitations resulting from episodic interaction.

2. We then discuss what is unique about connected relationships that could be used as part of a revenue model. We point to the increased value that is created by connected relationships; the higher dimensionality of the pricing mechanisms, which reflects more data availability; and the different timing of payments that result from the longer relationship.

3. We then link these idiosyncrasies of connected relationships to the revenue models and propose a framework to explore alternative revenue models.

4. Based on this framework, we articulate a set of guiding principles for choosing a revenue model and illustrate those with examples from other industries.

We conclude the chapter by discussing challenges related to privacy, an issue that is particularly relevant in revenue models in which customers pay the firm not with money but rather with data.

Revenue Models: A Brief Overview

Consider the history of pricing in four episodes. The first episode is haggling, still common at many bazaars. The vendor does not preannounce a price and haggles with each potential customer.

The second episode is posted prices, such as those printed on items in a supermarket, listed in catalogs of mail-order companies, or displayed on billboards. Posted prices simplify transactions, increasing convenience and efficiency. However, they enforce uniformity across customers. If Selena is willing to pay $500 for a phone but Jackson is only willing to pay $300, price discrimination between the two is difficult. Similarly, if a retailer has only one phone left in stock, it might make sense to raise the price, but this is often impossible if the price is posted.

With the arrival of the online marketplace, we entered a third episode. It became feasible to adjust prices dynamically and intelligently. As consumers, we are most familiar with, and sometimes annoyed by, airlines doing this. A flight from Philadelphia to Boston might go from $99 one day to over $400 the next, reflecting seat availability and the airlines' ability to identify us as a likely business traveler given our travel time. The internet also facilitates more complex pricing schemes, such as customer loyalty programs or group buying.

Yet despite these variations, from haggling at the bazaar to dynamic online pricing, traditional revenue models retain three limitations:

1. *Limited information:* Given the episodic nature of the traditional (nonconnected) business transaction, there comes a time when buyer and seller have to agree on a price, be it for a toothbrush or a medication. The problem is that the value the buyer will derive from that transaction is still unknown at that time. Will the new toothbrush really reduce my need for dental services?

2. *Limited trust:* One solution to dealing with limited information is to delay the final pricing decision until more information is

available. For example, the toothbrush manufacturer could re-quire the customer to pay another $500 if his teeth remain healthy. The problem with that solution is that in the case of a cavity, the customer will blame it on the toothbrush and the manufacturer will blame it on poor brushing behavior. Short of any monitoring data, the conflicting interests of buyer and sup-plier will erode trust between them.

3. ***Transactional friction:*** Even if we could overcome the limited trust and find a way to determine whether the customer's deg-radation of teeth is due to poor brushing or poor product per-formance, we still face the problem that the customer derives value from the toothbrush every day. Yet, traditionally, paying every day is a very costly practice. Every transaction requires an administrative overhead for payment processing, which likely will separate the customer's timing of payment from the timing of deriving value.

With advancements in connectivity and the resulting emergence of connected relationships as discussed throughout this book, we have now entered a fourth episode of pricing.

What Is New with Connected Strategies?

In this fourth episode of pricing, the three limitations just discussed are overcome by longer-lasting, connected relationships facilitated by high-bandwidth information exchange. It is possible to use a whole range of additional variables as part of the revenue model. In other words, the price can now depend on factors that previously could not be used to influence the pricing decision. This includes information about the following:

When the product was used

Where it was used

Who used it

What benefits were derived from using it

What problems occurred while using it

In short, the resulting revenue models can now be tailored to the particular use case. As a result, the connected relationship allows the firm to eliminate the three limitations discussed earlier in the following ways:

The problem of limited information can be overcome by delaying payments until more information becomes available. The most common revenue model that results from this is referred to as pay-for-performance. Payments are delayed until more information about the user benefits are known.

The problem of limited trust can be overcome as constant information flow allows for the monitoring of actions taken by two parties with conflicting interests. Such verification is also necessary for the pay-for-performance model.

Because of low transaction costs, there is no reason to lump all financial transactions into a single payment, as is common in episodic relationships. Instead, we can use revenue models such as pay-per-use (pay every time the product is used).

Thus, connected strategies allow us to create completely new revenue models. For our toothbrush, we can make the price a function of how many minutes the brush was used per day, how many different customers used the brush (hopefully with different brush heads), or the degree to which cavities could be avoided. In other words, the constant connection and associated information flow allow us to increase the dimensionality of the pricing space. No longer is there a single price printed on the box; there now exist many different options for revenue models, including different ones for different segments.

The increased dimensionality is appealing at first because it provides us with many levers that we can pull to increase profits. But, just as would be the case for anybody going from the driver's seat in a car to the pilot seat of an airplane, too many levers can be overwhelming. This raises the question, What are the general rules on how to form new revenue models, especially those that take advantage of connectivity? The following six guiding principles will help you answer this. Each is written as an action item and illustrated with our toothbrush example in the sections that follow, as well as with case studies from other industries:

1. Think value creation first.

2. Make pricing contingent on performance.

3. Remember the ecosystem is broader than the supply chain.

4. Get paid as value is created.

5. Reinvest some of the created value into the long-term relationship.

6. Be cautious when replacing cash payments with data payments from users.

Let's look at each principle in turn.

Principle 1: Think Value Creation First

Consider the payments received by ophthalmologists for performing eye exams on patients with diabetes—a practice generally recommended annually to prevent diabetes-related blindness. Patients don't particularly like these examinations because they require a time-consuming eye dilation for a retinal photograph to be interpreted by the ophthalmologist. The dilation can leave the patient with blurry vision for several hours. Patients who could otherwise get back and forth to the doctor on their own often need someone to take them home.

Adherence rates for exams are low, no doubt contributing to the high incidence of preventable blindness among diabetics.

A recent study showed that a commercial insurer would typically reimburse $254 for an in-office examination involving retinal photographs—about $26 for the photographs plus some facility fees and the professional service. The same insurer would reimburse a total of only $16 for the photographs when the service was performed remotely, with no payment allowed for the interpretation of the images.

This example shows that many business relationships are destroying value because of poorly aligned incentives. For example, in health care, the payer is not the consumer. Insurance companies, short of information, are concerned that patients consume too much and physicians provide too much. In the diabetes case, insurers apparently liked the higher price, pain, and friction of the traditional exams because they deter patients from having them. (Whether such preference would backfire because of higher costs down the road is a matter for debate.) Thus, they were comfortable with high reimbursements for an office visit but would only pay a small fraction for an equally effective remote service.

Before we think about how to build a revenue model for any business, we should ask ourselves what actions maximize the value in the system. Once we know the desired actions, we can think about a revenue model that rewards people for those actions. In the foregoing cases, we want diabetics to get their eyes examined and everyone to brush their teeth.

Rather than just replicating the old relationships—in the eye case, the insurers being concerned about fraud by doctors and overuse by patients—we should use connectivity to ensure that every actor makes value-maximizing decisions. For example, in the construction industry, contractors and customers are often at odds from an incentive perspective, where contractors get paid through costs-plus pricing, including an allocation for time worked. This incentivizes contractors to take more time to complete a project, harming the customer both by delaying the project and by charging a higher rate. Because customers often interact with contractors only once, the reputational loss of such behavior for contractors is low. This hurts both customers and the contractors who are actually doing a good job. Now, connected market makers such as Angie's List connect users via crowdsourced reviews to create transparency in a contractor's practices. Contractors who repeatedly exploit their customers will receive lower ratings, which translates into fewer opportunities. This aligns contractors' incentives more closely with the customers' incentives because their future business depends on their reputation.

Principle 2: Make Pricing Contingent on Performance

When deciding to buy a product, customers face uncertainty about how good that product or service will be. Whether they are consumers or firms, customers don't like uncertainty, and avoiding it can kill a value-creating transaction.

Pay-for-performance is one way to overcome the problem. Customers don't pay for the product or service; they pay for (some of) the value it creates for them. This is possible in a connected world because we have good data about the customer.

In the toothbrush example, connectivity allows us to observe the health of the teeth. This enables us to give the customer a form of performance guarantee ("If your teeth are not in good health, you will not pay a penny") while also aligning the customer's incentive ("If you do not brush, our performance guarantee no longer holds"), thereby avoiding the incentive inefficiencies mentioned earlier.

In general, we define pay-for-performance revenue models as revenue models in which fixed transaction prices are replaced by payments contingent on achieving certain objectives. The following examples help illustrate their use in a wide array of industries:

> As we noted in chapter 1, Rolls-Royce offers jet engines and accessory replacement services on a fixed-cost-per-flying-hour basis to airlines ("power-by-the-hour"), linking revenue with

performance. This is aided by onboard sensors to track on-wing performance.

Power purchase agreements (PPAs) are widely used in the solar industry. Rather than asking customers to buy the solar panels, pay for installation upfront, and wait for energy savings to trickle in, PPAs allow customers to lock in the energy produced by the installation at a fixed cost per kilowatt-hour over the term of the PPA, without owning the equipment or paying upfront.

Some consulting firms are moving from a billable-hour model to a fees-at-risk model, where a portion of the consulting fee is linked to their clients' results. This is in response to corporations' increasing desire for impact and outcomes over pure insights.

Principle 3: Remember the Ecosystem Is Broader Than the Supply Chain

As a manager of a business, you often think about your supply chain. You buy components, manufacture your product, and sell it to retailers who sell it to consumers. Unfortunately, there are only so many degrees of freedom in this supply chain—only so many parties you could go to for revenue. Or are there?

Recent research in strategic management has moved the focus from the supply chain to the ecosystem. The ecosystem is much broader and includes all firms or other organizational and individual entities that have some interest in your product. To figure out which entities are in your ecosystem, ask yourself, "Who else would derive value from our connected toothbrush?" For this product, the list might include entities such as the following:

Insurance companies paying for dental care

Dental practices that understand that the new brush means potentially fewer cavities and the need to be on the toothbrush's list for referrals

Toothpaste companies

Parents who are concerned about their children's brushing habits

Consumer product companies that would love to learn about the habits of consumers

Comcast, Verizon, and other carriers that are happy about almost anything that consumes bandwidth

Many of these companies would benefit if the Smart Connect XL3000 succeeded. In other words, they might be willing to share with us some of the value they get from our existence.

There have been numerous examples in which firms provided a connected product or service to their customers and made their revenue from other sources besides charging the customer. Examples include the following:

Many of the peer-to-peer network apps that we discussed in chapter 7 are free to customers. Some companies occupy two positions in the ecosystem: they are the organizer of the peer-to-peer network and the producer of a complementary product that is used in this network (example: Nike running shoes).

In the world of connected security, insurance companies subsidize the installation of advanced fire alarms. Similarly, car insurance companies offer discounts for drivers who are willing to have their driving monitored.

In the world of personal fitness, many gyms now get more revenue from insurance companies than directly from the users sweating on their treadmills.

Principle 4: Get Paid as Value Is Created

For most products or services, customers derive benefit over time. You buy a car and drive it for years. You buy running shoes and run five hundred miles in them. In the traditional episodic relationship, however, payment typically happens upfront in one chunk. You could send Nike fifty cents every time you go running, but it's likely neither you nor Nike wants to do that. More specifically, Nike would not trust you enough to give you a pair of shoes upfront without a payment guarantee in place. And you might not like the idea of sending Nike a payment every time you go running unless it was automated.

In a connected relationship these problems disappear, and the result is a victory over limited trust and frictional inefficiencies in payment. When your shoe talks to your phone and your phone connects to your bank account, you could pay Nike ten cents per mile. This is the pay-as-you-go model.

The world of hardware and software has seen a big shift from huge upfront transactions to pay-as-you-go models. Many firms no longer buy huge servers anymore, instead paying for "infrastructure as a service" from providers such as IBM or Amazon Web Services, paying on a per-use basis by the hour, week, or month. The key value for customers is a reduction of risk. Customers are never out of capacity, nor do customers have any idle capacity. They also have no upfront equipment or maintenance costs, which may be prohibitive for small companies that need cloud services. Another revenue model option is "platform as a service," which is priced per application or per gigabyte of memory consumed per hour. Platform-as-a-service providers include Google and Microsoft.

Similarly, software has moved in many cases from a model of purchasing and installing it on local machines to "software as a service," where customers pay based on features and use. Firms like Salesforce and Netsuite have adopted this revenue model.

Related to pay-as-you-go models are "freemium" models, where firms provide a basic model for free and charge for access to premium versions of their product. Dropbox and LinkedIn are examples. The free version attracts customers, while the premium version ("Ran out of storage space? Upgrade!") is used to drive revenue. Freemium models must find the right balance between giving away enough features for free to attract customers (especially when customer benefits increase with network size, as with LinkedIn) and retaining significant improvements in the premium versions to persuade at least a fraction of the customers to upgrade. Many newspapers and some magazines with an online presence have gone this route as well. A certain number of articles can be read for free each month, then customers must subscribe to access more content.

What makes many freemium models feasible is the ability to manage micropayments efficiently. With the advent of smartphone apps, in-app purchases have made the micropayment model seamless. Many apps start out as free for users, offering basic services or experiences before prompting the user to unlock premium content by spending a small sum. In China, for instance, Tencent introduced QQ Show, which allowed users to design their own avatar that could be used not only in the instant messenger of the QQ app but also for the group chat, gaming, and dating functions within the app. The customization options included appearance, virtual clothing, jewelry, and cosmetics. These items could also be purchased as a gift for other members. Each item only cost a few RMB (pennies in US currency) but created a significant revenue source for Tencent, given its more than eight hundred million active users on QQ. (For more on Tencent, see the sidebar.)

One of the biggest sectors for micropayments is video game development. Gamers buy virtual currency with real money and spend it to upgrade their characters, buy special weapons, access hidden levels, and speed progress in the game. While individual purchases can be small (as little as ninety-nine cents), aggregate purchases can be staggering. It has been estimated that the free mobile game *Clash of Clans* has earned more than $3.5 billion through in-app purchases (of products that practically have zero cost of production).

Micropayments also allow peer-to-peer networks to create payments across members. *Da shang*, or virtual tipping, is an increasingly popular form of micropayment for Chinese netizens. For websites or social media platforms that support this function, viewers can choose to

WECHAT: THE OPERATING SYSTEM FOR LIFE IN CHINA

WeChat, an app owned by the Chinese company Tencent, originally started out as a messaging system. It has evolved into an all-encompassing app that allows its users to have group chats, make calls, post personal news (including text, images, or videos), read news, order food, make doctor appointments, hail cabs, pay merchants, send money to friends, play games, and much more. Functionality is expanded by more than 580,000 mini programs that work similarly to separate apps but are housed within the WeChat app. WeChat now has more than nine hundred million daily active users spending more than an hour on the app on average. What makes Tencent as a company different from Google, Facebook, and other players is that most of its revenue stems from value-added services rather than from advertising. While Google derives more than 90 percent of its revenue from advertising, Tencent derives more than 80 percent of its revenue from micropayments for services such as in-app purchases within online games or fees for using the WeChat pay function.

virtually tip content creators when they are wowed by the experience. Places that have adopted this format include blogs, video sites, and various social media platforms such as WeChat. The model encourages content creators to put up quality content for free, hoping to recoup the development cost through tips.

Whether it's a freemium model or a micropayment method, they both capture one idea: get paid at the same time as your product or service creates value for your customers, because at that time, customers are often quite happy to pay.

Principle 5: Reinvest Some of the Created Value into the Long-Term Relationship

As we have seen, one great benefit of connected strategy is that you engage with your customers in a long-lasting relationship. From an economic perspective, that means the firm does not have to compete for every transaction with each customer. This translates into lower discounts and less spending on customer acquisition costs in sales and marketing. At the same time, value is also generated for the customers who no longer need to engage in costly and inconvenient searches and are provided with highly personalized offerings.

To create a sustainable advantage, it is important that at least some of the value that is created is reinvested, strengthening the repeat dimension of the connected strategy. Rather than taking the value and simply handing it back to the customer, as is done in traditional loyalty programs, the firm should seek to increase the level of customization it can provide. As we discussed in chapter 5, we can use the value created by connectivity to move further up the hierarchy of needs of our customers and establish our firm as a trusted partner.

At the level of a trusted partner, we are granted the authority of handling a broader need, be it oral care (the Smart Connect XL3000), education and career management (recall the example of Lynda.com), or wealth management. This responsibility is coupled with ongoing compensation, as illustrated by these examples:

> The original benefit of becoming an Amazon Prime member for a yearly fee was free two-day shipping on many items sold by Amazon. Over time, Amazon has reinvested and increased the benefits to include access to Prime Video (including licensed and original content), Prime Music, Prime Reading, photo storage, and other features. Each of these additional services increased customer value and the information that Amazon received about its customers, allowing it to further personalize its offerings and to increase customer loyalty. In

2018, Amazon Prime exceeded one hundred million members, counting half of all US households among them.

As a result of repeated interactions, subscription services can curate and contextualize based on the learned customer's preferences. Birchbox, a monthly beauty products subscription service, invests its created value in data and analytics to analyze what customers value most highly in order to better serve them with future products. This leads to lower churn rates, increasing the lifetime value of a customer and the ability to dedicate additional spending, which creates a positive feedback loop.

Principle 6: Be Cautious When Replacing Cash Payments with Data Payments from Users

Several of the most successful connected strategy companies have a seemingly odd revenue model: giving away their product. Google doesn't charge for searches or Gmail; Facebook and LinkedIn do not charge you to become part of their networks; and TripAdvisor does not charge you to find the most popular attractions in cities around the world. But obviously, nothing is free. Users of these sites do not pay with their money; they pay with their data.

Somebody searching for the term *spine surgery* on Google is very likely to have back pain. Spine surgery is a very profitable product line for hospitals and private practices alike, so knowing that Joe Miller in Chicago is looking for spine surgeries is something that health care providers are willing to pay for. How much? With the clearing price for most Google AdWords costing pennies per click, it is notable that those concerning medical needs currently stand at around forty dollars per click.

As this example shows, one key revenue stream can come from advertisers, who can use the data to create more targeted and more effective advertising campaigns. For instance, navigation apps such as

Waze do not make money by charging their users. Instead, they harvest user location data to display the most relevant location-based ads within the app. This determines what shops, restaurants, and other small businesses you see on the map as you drive.

Another key revenue stream can come from referral fees. For instance, Mint offers the convenience of managing all of a customer's personal finances in one place. While it is free to use, Mint generates revenue based on referrals made to consumer product companies or financial institutions that sell financial products or credit cards. It also derives revenues from the aggregation and distribution of user data. Although unique identifiers are removed to preserve individual confidentiality, the pool of real-time financial data has tremendous value in assessing consumer trends.

The almost irresistible psychological attraction of free products, coupled with the opacity of what data is collected and how it is used, often obscured in long terms-and-conditions statements—"click here to accept"—has led to a veritable gold rush of firms trying to collect as much data as possible through whatever means available. Firms often collect data with the sole purpose of reselling the information. It seems like the Wild West out there. To us, this development does not appear sustainable in the long run . . . and we would be very glad if that turned out to be the case. It is not hard to imagine that rising societal concerns over privacy, coupled with technological solutions that provide customers with much more control over their own data, will create a higher burden of proof for the feasibility of revenue models that are solely based on paying with data. We can imagine that in the future, privacy settings will be moderated and negotiated by customer-owned software that sits between the customer and the various data-gathering apps, rather than being hidden within the various apps. At that point, customers will have the ability to release their personal data slice by slice if they see true value in doing so. Until this technological solution is available, we can only caution firms that want to use the reselling of data as their main revenue stream. To create a truly sustainable revenue model will require firms to navigate a minefield that is constantly shifting.

First, as we have discussed before, the goal of a connected relationship is to become a trusted partner to the user, requiring a much higher degree of trust than necessary in a traditional episodic interaction between customer and firm. For this, we propose, a firm needs to help customers understand the price they pay, even if this is not a monetary price but rather only in the form of data. Paying through data can yield value to both parties, but it has to be transparent to the customers what happens with the data they provide.

Second, several technology experts recently have proposed to replace the pay-with-data revenue model with a pay-for-data model. The argument is that customers should not only be rewarded for their user-generated content, such as teaching Google how to recognize the human voice and collecting traffic data in the connected car by getting a free product or service, they should receive a cash compensation on top of this. Though the price of data ultimately should be shaped by market forces, the almost unlimited appetite for data of artificial intelligence–based businesses makes the idea of paying customers for their data at least an interesting twist to the revenue model. For instance, it has been estimated that Facebook's Instagram is worth some $100 billion and that its users have uploaded twenty billion photos. The following calculation is not meant to be scientific, but the numbers tell a story: a $100 billion valuation for twenty billion user-generated pictures—that equates to $5 per picture. Shouldn't those who took these pictures get a part of the pie? True, Instagram has done much more than just accumulate photos. Nevertheless, more and more technology experts have raised the question of to what extent users should be compensated for the data they provide.

Third, when customers pay with their wallet, jurisdictional questions—for example, tax implications of such transactions—are fairly clear. When customers pay with data, this becomes much more complex. Suddenly, questions such as where the data is processed and where it is stored matter a lot, as reflected by the 2018 implementation of the European Union's General Data Protection Regulation, which applies to all companies processing the personal data of customers residing in the European Union, regardless of the company's location.

As you are creating your connected strategy—whether selling data becomes part of your revenue model or data collection is used purely to extend your own relationship with your customers—you need to deal with these issues actively. Given how quickly this field is changing, you have to keep abreast of rapidly changing regulations and update your answers frequently.

Six Principles for Designing Revenue Models in a Connected Strategy

We opened this chapter by asking you about the right price for a connected product. Our discussion in the remainder of this chapter emphasized that creating a good revenue model is more than what is suggested by the word *pricing*. Instead, designing a revenue model is based on identifying the various players in the ecosystem, understanding their (typically conflicting) objectives, and leveraging technology, all with the objective of maximizing value.

Value is maximized when the corrosive forces of limited information, limited trust, and transactional friction are overcome, which plays to the strengths of a connected relationship. In this chapter, we have articulated a set of principles that will help guide you in the design of your own revenue model:

1. Think value creation first.

2. Make pricing contingent on performance.

3. Remember the ecosystem is broader than the supply chain.

4. Get paid as value is created.

5. Reinvest some of the created value into the long-term relationship.

6. Be cautious when replacing cash payments with data payments from users.

How can these principles be implemented? The workshop that follows the next chapter will provide you with exercises to help you use these principles to create the revenue model that's right for your business.

9

Technology Infrastructure for Connected Strategies

Advances in technology are critical to connected delivery models. How should we think about the enormous opportunities technology presents? Do you need to be a technology expert to create a connected strategy? This chapter aims to guide you through that maze. Rather than offering a comprehensive catalog of technologies that inevitably will date quickly, we provide a framework for thinking about connected technologies so that you can design and implement your connected strategy. We discuss specific technologies to illustrate general principles and illuminate what we hope is a more timeless perspective.

As in previous chapters, we choose a setting to illustrate key lessons. Our focus here is on home automation and the connected house. In three separate scenarios, Bill, Kayla, and Aruna are characters arriving at home in the evening, making themselves comfortable, and preparing dinner.

Bill is a college professor. He stops for groceries on the way home. On arrival, he puts his four bags down in front of the locked door, fumbles for his keys, and unlocks the door. The temperature in the house is on the cold side—the air conditioner has been running all day because Bill forgot to adjust it when he left that morning. Ignoring the granola crumbs on the floor, he walks to the coffee machine and turns it on, only to remember that he was out of coffee and forgot to put it on his grocery list. It will have to be tea today. While boiling the water, he catches the end of an NPR story he started listening to in the car—too bad he missed the middle part. Then he sits down on his couch with his tea. Afterward, he gives the floor a quick once-over with the vacuum.

Now consider Kayla, a high school senior living at home. As Kayla arrives, her mother opens the door and welcomes her. Mom, who came home from work an hour ago, has created a perfectly homey ambience for Kayla: the floor is clean, the air conditioner is set to seventy-three degrees, and the smell of fresh coffee is in the air. Mom went grocery shopping on her commute home, so the fridge and pantry are fully stocked, even though Kayla and her friends depleted the snacks while watching movies the night before. From the restocked fridge, Kayla grabs a diet soda and sits down in front of the television to watch her favorite sitcom while waiting for Mom to complete dinner. (Just for the record, Kayla's dad might equally have done the shopping, cooking and cleaning, and we pause here to salute everyone who runs a home with children in it.)

Finally, consider Aruna, a tech executive. As she arrives home, her front door unlocks without her touching it, opening promptly based on a sensor in the door recognizing her identity. She enters the house and is happy to see that the vacuum robot, Roomba, has completed its job. The temperature is set at a comfortable seventy-five degrees, thanks to the preprogrammed Nest thermostat. Aruna shouts out, "Alexa, turn on the coffee machine," then heads over to the pantry, which is fully stocked thanks to Amazon's home delivery services. Aruna grabs a drink and sits down. Like Bill, she had been listening to something in the car. In her case, it was a podcast, and it restarts exactly where she had left off.

We hope that at least at some point in your life you have been spoiled by a parent, friend, or spouse, just like Kayla. We assume that you are old enough to remember air conditioners controlled by a gray box on the wall and that you have done the chores of managing a household, including vacuuming and grocery shopping. And we assume you have at least heard of the products present in Aruna's household: iRobot's Roomba, Amazon's Alexa, Google's Nest thermostat, and the smart home security system that detects who is at the door.

Here is a preview of how we will use the three user experiences in the remainder of this chapter:

> Just like other connected customer relationships, the connected home user experience comprises many individual pieces. Alexa, Nest, and Roomba perform specific functions, like making coffee, regulating the temperature, and vacuuming. Technologies to clean the floor include automated vacuum machines (Roomba), traditional vacuum machines, and somebody with a mop and broom. In the second section of this chapter, we talk about how to deconstruct a connected strategy into a set of functions that each represents a job to be done.

> Once we know what functions the technology should perform, we can think about the technical means to accomplish them. But, as we go into the implementation, as managers we always should remember that users derive value from what a device does, not from its underlying technology. In the third section, we will describe the concept of a technology stack, with the functions as the user sees them on top of the stack and the technical details at the bottom.

> Implementing a function to support a connected customer relationship has many design options. The brewing of coffee can be triggered by voice activation via Alexa, through an app on our phone, by sensing our proximity to our home, or through a traditional on/off switch. We should explore many alternatives and get inspired by solutions from other industries. In the fourth section, we will discuss the classification tree as a powerful tool

to explore design options and selection tables to help us choose between them.

Ten years ago, Aruna's scenario would have sounded futuristic. At that time, very few people could afford such whiz-bang technology for home use; today this scenario (or at least parts of it) is widely attainable. The change has been wrought by improvements in technology that provide more functionality at lower cost. As we discuss in the fifth section, advancements deep down the technology stack bubble up and enable new connected relationships that were previously impossible or prohibitively expensive.

Deconstruction: Breaking Down a Connected Strategy into a Set of Functions

Technologies do not have value per se; users derive value from the technology performing a specific function. We can think of a function as the purpose of the technology. The purpose of the technology answers the *what* question (what does the technology do?). In our example, Kayla does not care whether it is a smart door with face recognition and an automatic lock that lets her in or her mother who does it. In the same way, the cleanliness of the house is what matters, whether achieved through Roomba or a member of the household with a broom or vacuum cleaner.

Once we are clear about the *what*, we can turn our attention to the *how* (how does the technology work?). Functions are performed by devices, pieces of software, or people following a workflow. Sometimes the functions are performed by the customer themselves, as was the case when Bill did his own grocery shopping and virtually everything else in his scenario.

The first thing you should do when exposed to a set of technology buzzwords in the context of connected technology is to forget about the *how* and focus on the *what*. In any connected relationship, there are many whats, so we need to focus on something more specific. We focus by deconstructing the connected strategy into a set of required sub-functions. Deconstructing a problem means breaking it up into smaller, manageable subproblems and solving those first.

We find it helpful to deconstruct a connected strategy based on two dimensions. The first dimension captures all the functions that need to be carried out within the two building blocks of a connected strategy: the connected customer relationship and the connected delivery model. In this first dimension, as we saw in chapters 4 and 5, the connected customer relationship consists of four pieces:

> *Recognize* concerns becoming aware of the need in the first place.

> *Request* includes the search and decision-making process, the placement of the order, and the processing of the payment.

> *Respond* captures those functions required so that the customer can receive the product or service, experience it, and be connected to any form of after-sales support.

> *Repeat* encapsulates all the functions that allow the firm to learn continually from the repeated interactions it has with its customers.

Within the connected delivery model, as described in chapters 7 and 8, we need the following:

> The functions required to establish and support the connection architecture, which means the connections among the firm and its suppliers and ecosystem. For example, this could be a link to a supplier for a connected retailer or a reputation scoring of an individual within a peer-to-peer network.

> The functions required to establish and support the chosen revenue model. This could include measuring use time, assessing

the performance of the product, or transmitting data to other ecosystem members.

The second dimension of deconstruction takes each function, such as identifying a person, making a payment, or shipping a good, and breaks it up further into four types of subfunctions: **sensing**, **transmitting**, **analyzing**, and **reacting**.

Why those four? To illustrate, let's return to our connected home scenario, this time focusing on the thermostat. To avoid excessive air conditioning, which is both uncomfortable and wasteful, four functions need to be completed. The current temperature needs to be *sensed*, it needs to be *transmitted* from the sensor to a decision-making unit, that unit *analyzes* the information and makes a decision, and then somebody or some device needs to *react* by executing the decision. This creates a feedback loop common to all connected technologies: sense-transmit-analyze-react, with the acronym STAR.

We can now combine the two dimensions into a table, as is shown in table 9-1. The columns capture the different elements of the connected strategy: recognize, request, respond, and repeat, plus the connection architecture and revenue model. The rows capture the four dimensions of STAR: sense, transmit, analyze, and react. We can use the table to catalog the many subfunctions needed to create a connected strategy, as table 9-1 shows for Aruna's coffee consumption.

Consider the first column, recognize. One subtask is to sense that Aruna has only twenty grams of coffee left. This quantity information then has to be transmitted to a cloud or edge computer (a computing device sitting close to the information source). There it must be analyzed to answer the question whether the amount left is less than the desired minimum quantity of fifty grams. Lastly, the system must react and start the request module (next column) to reorder the coffee.

Or consider the last column, revenue model. Assume the coffeemaker didn't charge Aruna up front for the machine but charges her a daily fee that includes a guarantee that the machine will have an uptime of 100 percent. To implement this revenue model, it is imperative to sense the maintenance needs of the coffeemaker on a continuous basis. This status information must be transmitted to the coffee machine's service provider. This information is then analyzed and a decision made about when to replace the machine. And lastly, the firm must react and ship a replacement machine once the old one shows signs of wear.

As we have shown in table 9-1, each function of a connected strategy can be broken down into further subfunctions using the STAR approach. At the end of this deconstruction, you have a set of very specific subfunctions. Each subfunction, in turn, corresponds to an engineering problem. In other words, you have a job to be done and now can look at available technologies to perform it effectively. (For another example of the STAR approach, see the sidebar.)

Functions Are Performed by a Technology Stack

Returning to Aruna's connected home, let's look at what needs to be in place for the door to open conveniently on her arrival. We need a camera at her door, transmission technology to send the video stream, and a computing device that takes the incoming data and sends a signal to a locking mechanism that opens the door or keeps it shut. We can identify people by capturing their biometrics (facial images, fingerprints, eye scans), by having them enter a user ID and password, or by sensing the proximity of a device such as a key or a phone. If we drill down on face recognition, options include 2-D and 3-D image processing. Within 3-D image processing, we can further distinguish methods of face recognition that rely on unsupervised deep learning methodologies using neural networks and other methods that work based on the predefined geometric patterns of faces.

Chances are that you are not that interested in the nuts and bolts of unsupervised deep learning methodologies using neural networks. You just want Aruna's door to open when she arrives but remain closed

TABLE 9-1

Two dimensions for deconstructing a connected strategy

	Recognize		Request		Respond			Repeat	Connection architecture	Revenue model
	Become aware of the need	**Search and decide on option**	**Order**	**Pay**	**Receive**	**Experience**	**After sale**	**Learn and improve**	**Connect parties in ecosystem**	**Monetize customer relationship**
Sense	Notice the amount of coffee left in the pantry	Receive information about current coffee prices from nearby retailers	Make sure the desired item is available	Check current cash balance in bank account	Notice arrival of delivery	Notice home owner approaching the front door	Measure change of heart rate and pupil dilation after first sip of coffee	Identify the user whenever she drinks coffee, regardless of location	Sense ordering needs of nearby homes	Check the functionality of the coffeemaker
Transmit	Send quantity information to computing system	Get those prices onto the central system	Send order to retailer	Combine account status with projected grocery orders	Send identity information to central system	Send arrival information from door to server	Send information to edge computing system	Send identity and preference information to central system	Pool neighborhood information to one server	Send status report about coffeemaker to service provider

TABLE 9-1 (continued)

	Recognize		Request		Respond			Repeat	Connection architecture	Revenue model
	Become aware of the need	Search and decide on option	Order	Pay	Receive	Experience	After sale	Learn and improve	Connect parties in ecosystem	Monetize customer relationship
Analyze	Compare with target quantities	Look at prices and volume discounts, potentially factoring in travel plans	If desired item is not available from preferred source, find best alternative	Look for potential account overruns and possible loyalty rewards	Make sure that delivery has been authorized	Identify the person in front of the door	Assess delight of user with coffee brand using physiological measurements	Analyze coffee preferences—e.g., depending on time of day	Evaluate eligibility for group discounts	Make decision when to send coffee-maker replacement
React	Decide it is time to start the request module for reordering	Activate order module	Place order for a particular item and provide shipping address	Execute payment	Provide access to pantry or coffee machine	Open the door and activate coffee brewing	Feed results into repeat module	Feed results into request module and help coffee roasters to improve their product	Negotiate special price with retailer	Send coffee machine replacement when necessary

Note: Each cell corresponds to a specific subfunction in the coffee-brewing scenario. The table is best read column by column.

APPLYING THE STAR APPROACH TO A SCHIZOPHRENIA DRUG

As briefly noted in chapter 8, in 2017 the FDA approved the first drug to be paired with a digital ingestion-tracking system aimed at improving patient adherence with taking medications. Medication adherence is a major challenge for some schizophrenia patients, as well as for those suffering from other conditions. The system senses when the pill is swallowed and then transmits the data. The drug Abilify is part of a drug-device combination branded as Abilify MyCite.

Consider the recognize dimension that is involved in the connected strategy of Abilify MyCite using the STAR approach discussed in this chapter:

- **Sense:** Embedded in each pill is a sensor (known as an ingestible event marker), which is the size of a grain of sand. The sensor reacts when it reaches fluids in the digestive system.

- **Transmit:** The ingestible event marker transmits a signal to a patch worn by the patient, which in turn is transmitted to the patient's phone, and from there to a cloud-based server.

- **Analyze:** The company's software compares the events associated with medication intake to a medication regimen established by the care team.

- **React:** In case of a significant discrepancy, the next step of the connected strategy—request—is triggered, alerting the patient, family, or care team to take corrective action.

to everyone else. To delineate the underlying technologies, their functioning, and the business services they perform, it is helpful to think of technologies in the form of a stack consisting of hierarchical layers.

The most technical layers are at the bottom of the stack. For instance, the lowest level might be about the physical transmission of bits from one device to another. At this layer, you are talking about volts or frequency and worry about signal strength or network topology. The next layer up in the stack takes these capabilities as given. You know that bits somehow will get from one device to the other, so you can turn your attention to creating connections, which might involve establishing and ending connections between two devices through a protocol. The next layer might be concerned with sending data packages through a network using sender and receiver addresses, and so on. At the top of the stack is the application layer, which is closest to the end user.

The beauty of any stack model is that the user can ignore the lower layers in the stack, just as you can drive a car without knowing how a combustion engine works. Stacks create clear interfaces and layers of abstraction. As somebody building connected strategies, you can decide for yourself how deep down into the stack you want or need to go.

The collaboration between Steve Jobs and Steve Wozniak illustrates how much you need to know about the lower layers of the technology stack (and how far you can advance in your career by excelling at the top layers). In the early days of Apple, Jobs was the visionary imagining user experiences. He was primarily concerned about the higher layers in the stack, while Wozniak was the engineer who made it happen, which required him to dive into all the technical details lower in the stack. The focus on the user experience and the willingness to abstract from engineering details at lower levels stayed with Jobs throughout his career, including through the development of iconic products such as the iPod, the iPhone, and the iPad. Don't get us wrong: to make a connected strategy happen, somebody has to go deep down into the technology stack, but that somebody might not have to be you.

Functions Can Be Carried Out by Alternative Technologies

As you go down the stack, moving from user experience into technical details, you have design options. There almost always exists a set of alternatives to implement a function. Because this is not an engineering book, our focus is on the application level of the stack ("How can we recognize a person?"), though the logic applies to any level ("How can I transfer ten megabits per second over a distance of five meters?").

Let's look at a subfunction from table 9-1 and think about alternatives for implementing this function. Again, let's pick the subfunction "identify a person" and systematically explore our design options. A great tool to advance that exploration is the classification tree in figure 9-1.

A classification tree takes the space of all possible solutions and breaks them up into different categories. When it comes to recognizing a person, we can, at the highest level, distinguish between human solutions (a doorman or Kayla's mother) and automated solutions. Automated solutions can be further broken up into those that require user action and those that don't. And so on . . .

The classification tree helps you be systematic in your exploration of technological options and your discussions with engineers.

To populate the classification tree, we find it helpful not only to generate alternatives internally but also to look at how a subfunction is performed by other firms, especially firms outside your own industry. Consider the following example: While writing this book, one of our friends had a BMW that required maintenance. When the maintenance was completed, she was called by the BMW dealership and notified that the car was ready for pickup. She took a cab to the dealership,

FIGURE 9-1

Classification tree for the subfunction "identify a person"

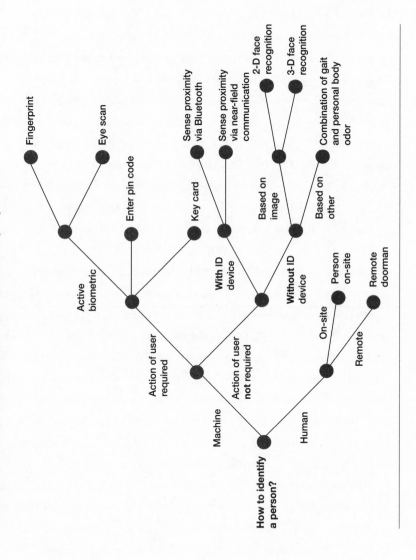

checked in, and was informed that the car was indeed ready and an employee would retrieve it from the off-site parking lot. Fifteen minutes later, the car arrived.

What other alternatives exist to implement the subfunction we could call "product handoff"? Let's look beyond car dealerships and contrast BMW's implementation with how Wawa, a convenience store chain in the mid-Atlantic region known for great made-to-order food, performs basically the same subfunction. At Wawa, you can order a sandwich through an app and then pick it up at the store (a respond-to-desire customer relationship, as we discussed in chapter 4). Since you want your meal to be fresh when you pick it up, Wawa uses geo-fencing with the customer's phone to sense proximity and have the food ready on arrival. Preparing it just in time creates a magical user experience. With this kind of service, a regional convenience store selling $5 sandwiches outperforms a global automaker selling $50,000 cars.

To further emphasize the power of this approach, recall our Disney example from the beginning of the book. Disney did not create the connected bracelet. Instead, Disney found this technology in hospital settings. The lesson is that, in almost all cases, companies already exist that have done a great job implementing a particular subfunction. Your job is to learn from the best, not to reinvent the wheel. This simple yet powerful observation also gets us back to Steve Jobs. He did not invent the graphical user interface when he and Wozniak launched the Mac. This function was created by Xerox PARC, which is where Jobs first saw it and, quickly scaling his mental classification tree, realized what could be done with it.

The output of the classification tree is a list of design alternatives. This list can be summarized in a selection table, as is shown in table 9-2. The selection table takes the design options as its rows and compares them along a set of dimensions—in our case, performance, cost, existing applications, and other comments. For illustration, we further divided performance into the subdimensions of user convenience and safety or reliability. Each option is then rated relative to the other options along each dimension.

TABLE 9-2

Selection of alternatives for "identify person" subfunction

Technology	Convenience	Safety/ reliability	Cost	Applied in . . .	Comments
Human at door	++	++	– –	Hotels	
Human remote	+	+	–	Building access	
Key card	–	++	+	Hospitals	
Enter pin code	–	–	++	Gym	
Fingerprint	–	++	+	Phone	
Eye scan	– –	++	–	Global border entry	
Sense device proximity using near-field communication	++	+	+	Car keys	
Sense device proximity using Bluetooth	++	+	+	Wawa	
2-D face recognition	++	+	+	Hotels	
3-D face recognition	++	++	+	High-end phone	
Combination of step pattern and smell	++	?	– –	Security dog	Technically not yet feasible

Notes: ++ = very good; + = good; – = poor; – – = very poor; ? = unknown.

Bottom-Up Innovation by Moving up the Stack

It is interesting to see how a stack changes over time. Consider Amazon's Alexa, which was helping Aruna prepare coffee. At the top of the stack, the level of the user experience, sits a function we could label voice recognition. Voice recognition is *what* voice recognition software does, but *how* it does this is a matter for the lower levels of the stack.

Looking only at the stack's top layer, one might say, "Voice recognition has been around forever," a valid statement since companies like Bell Labs and IBM experimented with it over a half century ago. For example, at the 1962 World's Fair in Seattle, IBM revealed a device called the IBM Shoebox, a computer the size of (you guessed it) a shoebox that had a revolutionary capability. The device had ten small lamps and a microphone. If somebody said "seven," then lamp number 7 would light up. If somebody said "four," lamp number 4 would light up, and so on. Engineers envisioned that we could soon dial a telephone number through voice commands.

Voice recognition was further improved with advances in computing technology. In the 1980s, a new approach to voice recognition emerged using a method known as hidden Markov chains. This technology would not simply listen to the sound and then try to match the sound with a word from a library, it would also factor in the probability of the word occurring by analyzing the words preceding it. If the previous word was *grand*, it is more likely that the following word will be *son* rather than *sun*.

The first mass-market application of voice recognition came with the software Dragon Dictate, a product initially retailing for some $9,000 and requiring substantial training to acclimate to the voice of a user. Dragon's software improved throughout the 1990s to require less training and sold at lower costs.

As computers gained increased processing power, voice recognition was built into more applications. You might be surprised to learn that both Microsoft Windows and Apple's Mac OS had voice recognition built in by the early 2000s. If you had a computer back then, most likely you did not use that feature because it was practically useless—unreliable and slower than just moving your mouse. So, voice recognition remained a niche market, constrained by its limited accuracy.

The breakthrough came in 2010 when Google added voice search to Android phones and Siri appeared as an iOS app the same year.

Enabled by the internet, Google and Apple captured the voices of millions of users who performed billions of queries, every one of them adding to their library of spoken words, making voice recognition what it is today.

The story of voice recognition demonstrates how improved functionality, greater accuracy, less training, and lower cost are a result of new technologies happening deep down in the stack. Hidden Markov chains, more processing power, and the internet—none of these technological advances was related to voice recognition. But as new technologies become available that impact layers deep down in the stack, they improve the execution of a function at the technical level. This improvement then bubbles up through the layers of the stack, enabling better functionality at lower costs in the application layer at the top of the stack. An example is the recent success of electric cars. Enabled by better battery technology deep down in the stack, electric cars now have become competitive with combustion engines. The driving does not change, but the user gets faster acceleration and a lower carbon footprint because of the advances in technology.

Technologies bubbling up in the stack and enabling new functionality at the application layer have an important consequence for the design and origin of your connected strategy. So far, we have described a top-down process: you create a vision for a connected strategy, you deconstruct it into subfunctions, and you look for technical solutions for each subfunction. A complementary way to create new connected strategies is from the bottom up. You spot a new technology and ask yourself, "Which application will be significantly improved if I introduce this technology to the stack?" In these innovation efforts, we start with the *how*, look for substantial advances in performance or cost reductions, and then ask the modified *what* question: "What new user experiences are enabled by this technological advance?" For two business-to-business examples, see the two sidebars on digital twins and drone delivery.

DIGITAL TWINS

The idea behind a digital twin is deceptively simple: create a digital replica of a physical system, be that an airplane engine, a wind turbine, or a piece of critical equipment on an oil rig. Digital twins can be used in product development (e.g., exploring the best shape for a wind turbine blade), product manufacturing (e.g., simulating different manufacturing processes to build the blades), and product operations (e.g., assessing whether the performance of the turbine is degrading prematurely). Digital twins are an important tool for product-lifetime management, creating a so-called digital thread for the product. The idea of digital twins is not new, but vast improvements in sensors, the power of edge computing, data transmission, and data analysis via artificial intelligence have allowed digital twins to become a truly valuable and cost-effective tool. This is an interesting example of technologies bubbling up in the stack in order to make higher-level connected strategies possible.

For instance, German machine tool manufacturer Heller used underlying data-capture technology and Internet of Things software by Siemens to develop a system that allows its customers to receive information about the performance of their tools in real time and simulate different maintenance strategies. This information allows Heller to provide preventative maintenance or, if there is an unexpected machine overload, to intervene immediately, reducing machine downtime for its customers. Moreover, the availability of this information has allowed Heller to innovate with respect to its revenue model. For customers who previously could not afford the purchase of a machine tool, Heller now offers a new machine usage model: it provides its machines to its customers, ensures 24/7 uptime, and charges for the time the machine is running.

DRONE DELIVERY

As managers, in addition to monitoring how other companies perform specific subfunctions in our classification tree (recall our discussion of BMW vs. Wawa), our job is to monitor the environment and look for promising new technologies that have the potential to bubble up through the stack. Unmanned aerial vehicles have a long history in military settings, including both world wars. As the technology advanced, performance improved dramatically. These devices, now called drones, were made in much smaller versions for espionage and in much larger versions for launching missiles, such as the Hellfire air-to-ground missile. With further advances in technology, drones soon became even smaller, more reliable, and cheaper. In 2018, the German company Volocopter announced an agreement with the government of Dubai to launch a taxi service relying on autonomous flying cars in the form of a multirotor aircraft. At roughly the same time, internet giant Amazon filed a string of patents for delivery drones, indicating that drones have emerged as a potential substitute for UPS and FedEx trucks. This is especially true in areas where traditional delivery trucks are hard to operate, as illustrated by the startup Zipline, which is experimenting with drone delivery for replenishing medication to hospitals in rural Africa. As managers, we need to be continuously on the lookout for technologies from many arenas, as Jobs did with Xerox PARC.

Innovative Business Models Don't Necessarily Require New Technologies

The key lesson of this chapter is that you can create innovative connected strategies without being a technology expert. You start with the business vision and then work from there, including deconstructing the customer journey into a set of subfunctions and then broadly exploring alternatives for each of them using classification trees. Alternatively, you may notice that technological advances have occurred that enable you to solve subfunctions in a better or more cost-effective way, allowing you to create new connected strategies.

In the process of populating the classification trees, you want to look for best practices for a particular subfunction both inside and outside your industry. Just cherry-pick what works well: pick a "sensing person arrival" subfunction from the hotel industry; pick "3-D face recognition" from a phone security system; and pick "execute payment" from a peer-to-peer payment platform. Then combine those pieces as you envision your own connected strategy.

You might view this approach of primarily relying on existing technological solutions as lacking in vision—where's the originality? It seems somewhat risk averse, shying away from technological breakthroughs. We propose that this approach is a strength, not a weakness. First, the originality is often in the *use* of the technology, not the technology itself. Ride-hailing companies did not develop GPS, cell phones, or Google Maps, but using these technologies allowed them to create a new connected strategy. Second, technology is improving so quickly that new solutions arise all the time. Thus, the best implementation of a particular subfunction is always a moving target. And third, relying on existing technologies can reduce risk. The history books on technology are full of failures. Many of them were costly, especially in those cases in which firms tried to deliver too much too soon.

Also, notice that our framework allows for more visionary thinking than might appear to be the case initially. When Amazon announced its potential usage of drones around 2016, the business world was in awe. As we show in the sidebar, delivery of consumer packages by drone might indeed be a novel way to implement the subfunction "rapid delivery." It might seem original; however, the military has long relied on drone technology, and drones have been used to deliver medication replenishment in rural Africa. A novel solution can come from the application of existing technologies in new settings. It does not always require fundamentally new technologies.

Finally, if you desire a visionary connected relationship that relies on yet-to-be-invented technology, we advise you to check out your local library's science fiction section. Sound crazy? Consider a detail in the history of the MagicBand we have not shared so far. We mentioned that Disney executives were inspired by hospitals where dementia patients were tracked using bracelets. But from where did the hospital executives get the idea?

The story of tracking people goes back to an old Spiderman comic strip from the 1960s in which Spiderman is forced by his evil enemy, Kingpin, to wear what is described as an "oversized ID bracelet in the form of an electronic radar device." In 1983, Judge Jack Love of Albuquerque, New Mexico, initiated the first judicially sanctioned use of monitoring devices. Inspired by the comic strip, Love envisioned defendants under house arrest or on parole wearing an ankle monitor that would signal its location. The world of science fiction has a long and remarkable track record of envisioning new functions before even the first technical applications exist. So keep on reading those comics and don't miss the next James Bond movie.

10

Workshop 3

Building Your Connected
Delivery Model

In this workshop, you will apply the frameworks we introduced in the last three chapters to help you build your own connected delivery model. Following the flow of the last chapters, you will do this in three parts. In part I, you will map your connection architecture and the connected customer experiences you create and compare them to your competitors' by using the connected strategy matrix. Part II will help you think about your revenue model, i.e., how you capture some of the value that you create. Finally, in part III you will catalog the key technologies that are part of your technology infrastructure.

Once you have captured the status quo, each of the three parts will challenge you to rethink what you are currently doing and help you take advantage of a connected strategy:

- **Part I: Building Your Connection Architecture**

  - *Step 1:* Use the connected strategy matrix to map your own activities and the activities of your competitors.

- ***Step 2:*** Use the empty cells in the connected strategy matrix to create new ideas for connected strategies.

- **Part II: Creating Your Revenue Model**

- ***Step 3:*** Understand your existing revenue model, identify its main limitations, and consider alternatives for your current activities, as well as for the ideas created in the previous step.

- **Part III: Choosing Your Technology Infrastructure**

- ***Step 4:*** Deconstruct your connected strategy into technological subfunctions and then catalog currently used technological solutions for each subfunction.

- ***Step 5:*** Identify new technological solutions and how those might enable further innovations in your connected strategy not identified so far.

Part I: Building Your Connection Architecture

Step 1: Use the Connected Strategy Matrix to Map Your Own Activities and the Activities of Your Competitors

In this step, we want you to use the connected strategy matrix to capture the status quo in your industry. You will do this by comparing your connection architectures and the connected customer experiences they create with what your competitors do, including longtime competitors as well as recent entrants to your industry. In the next step, you will use the connected strategy matrix as an innovation tool.

For this first step, start by making a list of your competitors, old and new alike. Next, turn to the matrix in worksheet 10-1 and use it

The connected strategy matrix: Inventory of your and your competitors' actions

	Connected producer	Connected retailer	Connected market maker	Crowd orchestrator	P2P network creator
Respond-to-desire					
Curated offering					
Coach behavior					
Automatic execution					

to position your own activities, as well as those by your competitors. What customer experiences are you and your competitors creating? What connection architectures are used for that?

It is often particularly insightful to understand where new entrants are emerging. By mapping them onto the connected strategy matrix, you can see what kind of strategy they are using and how they differ from existing competitors. This might help you spot where you are vulnerable for disruptions.

Your first task in this workshop is to fill out worksheet 10-1. Remember, a firm can be in more than one of the cells in the matrix. Many firms try to create more than one connected customer experience. Likewise, while most firms operate only in one column, some firms (e.g., Netflix, Amazon, Nike) operate with multiple connection architectures, as we saw in chapter 7.

Step 2: Use the Empty Cells in the Connected Strategy Matrix to Create New Ideas for Connected Strategies

Giving managers a blank sheet of paper and imploring them to innovate is often a frustrating experience for everyone involved. Where do you even start? We find that the connected strategy matrix can serve as a helpful tool to guide your innovation process.

For each cell of the connected strategy matrix, especially for those cells in which you are currently not active, ask yourself the following questions:

- What would it mean if we tried to operate in this cell?

- What kind of service or product would we offer to our customers?

- Which of the necessary activities would we engage in ourselves, and which would we provide through other players in the industry?

- What kind of connections to other players would we have to create to pull this off?

This exercise forces you to stretch your thinking (and your existing business model), particularly for those cells that lie outside your existing column—that is, your current connection architecture.

A second starting point for idea generation is to ask which inefficiencies could be reduced or what better services could be created if you could connect (currently unconnected) entity A with entity B. As we saw with the examples of grocery retailing and ride hailing in chapter 2 and then again in our discussion of connection architectures in chapter 7, efficiency can be improved by forming new connections. So, as you consider the different connection architectures, ask yourself the following questions:

- What are the most expensive resources in creating our product or service now?

- Where do we waste costs or capacity now, and how might this be reduced if we connected entity A with entity B?

- How could a connection architecture be used to share or reduce the risks associated with fluctuating demand or other forms of volatility?

For instance, in the hospitality business, the most valuable resource for a connected producer is a hotel room. Capacity gets wasted when a room is idle, either because it was not booked in the first place or because of a last-minute cancellation. By connecting empty rooms with potential guests—as (connected market maker) Expedia or (crowd orchestrator) Airbnb does—new reservations are made and empty hotel rooms are reduced. In addition, peer-to-peer networks and crowd orchestrators can help provide additional capacity when demand is high.

As you are considering new connection architectures or connected customer experiences, be it to increase the willingness-to-pay or to improve efficiency, you need to answer the following questions:

- What information flow will have to happen between these players? (This foreshadows the technological infrastructure that you might need to put into place, which you will analyze in more depth later in this workshop.)

- What incentives are you going to provide to the various players to be connected by you? (This foreshadows the revenue model that you will have to adopt, the subject of the next part of this workshop.)

Use worksheet 10-2 to keep track of your work.

Part II: Creating Your Revenue Model

Step 3: Understand Your Existing Revenue Model, Identify Its Main Limitations, and Consider Alternatives for Your Current Activities, as well as for the Ideas Created in the Previous Step

How do you make money? First, think about all the financial flows that you have with your customers. The following questions might help with this:

- What does the customer pay for?

- What are your different revenue streams? (e.g., initial sales, service)?

- Who is paying (e.g., the user, a third party)?

- When does payment occur? (At time of purchase? At time of use? Once or over time as in a subscription model?)

Next, look for inefficiencies in your revenue model. Do you use this revenue model because you believe it is the right one, or are you constrained by connectivity to the customer? Recall from our discussion in chapter 8 that there are three typical inefficiencies:

The connected strategy matrix: Imagine your company in each cell

	Connected producer	Connected retailer	Connected market maker	Crowd orchestrator	P2P network creator
Respond-to-desire					
Curated offering					
Coach behavior					
Automatic execution					

- Limited information flow

- Limited trust

- Transactional friction

Now that you understand the current revenue model, consider ways to overcome these inefficiencies. For example, in chapter 8, we discussed the following improvements:

- Reduce inefficiencies due to a prior lack of monitoring ability.

- Make pricing contingent on performance.

- Get paid as value is created for the customer.

Moreover, ask yourself who, besides the customer, benefits from your product and who in the ecosystem would benefit from the data that you are collecting. Through the right choice of revenue model, you might be able to benefit from value that is being created in the entire ecosystem, not just from your customer.

Finally, turn to the new ideas you created in the previous steps and look for revenue models for them as well. What new revenue model would you implement if you were able to create new connected customer experiences (e.g., coach behavior or automated execution)? In other words, what new revenue models would be possible or required if you decided to move to new rows in the connected strategy matrix given your existing connection architecture?

In the same manner, you also need to think about how changing your connection architecture would influence your revenue model.

Part III: Choosing Your Technology Infrastructure

Step 4: Deconstruct Your Connected Strategy into Technological Subfunctions and Then Catalog Currently Used Technological Solutions for Each Subfunction

By this point, you have hopefully created a list of possible ideas for connected strategies for your organization. Refer back to the ideas you came up with in the workshop in chapter 6 and the ideas you came up with in parts I and II in this workshop.

Next, we want you to deconstruct your connected strategy ideas by identifying the necessary technological functions that need to be carried out. As we discussed in chapter 9, this is best done along the steps of the customer journey and the additional dimensions of connected architecture and revenue model. Each of those then can be further deconstructed into subfunctions using the STAR approach, asking for each function what is needed along the dimensions of sensing, transmitting, analyzing, and reacting. Keep track of your analysis on worksheet 10-3.

Each cell in worksheet 10-3 corresponds to a "job to be done." List what possible technological solutions exist to complete each of these jobs in worksheet 10-4. Note that the same technology might solve several subfunctions. This step should not only involve an internal search of technology options that you get from your own technology experts, but also your observations of how other firms are solving this particular subfunction. As a starting point, here are some technologies, grouped using the STAR typology:

SENSING TECHNOLOGIES

In this category fit all technologies that directly measure aspects of the world that hold clues about the needs or desires of customers or that help users express their needs. This category includes all types of

Deconstruct your connected strategy into subfunctions

| | Recognize | Request | | | Respond | | | Repeat | Connection architecture | Revenue model |
	Become aware of the need	Search and decide on option	Order	Pay	Receive	Experience	After sale	Learn and improve	Connect parties in ecosystem	Monetize customer relationship
Sense										
Transmit										
Analyze										
React										

sensors, whether they are embedded in devices or in roads or are wearable or ingestible. This category also includes technologies such as gesture and voice interfaces, and conversational platforms that make it easier for customers to express their needs (and that ask for clarification if the need is not completely understood). Likewise, augmented and virtual reality technologies enable customers to express and understand their needs and desires—for instance, by showing customers different options in a very life-like setting.

TRANSMITTING TECHNOLOGIES

The ubiquity of high-speed internet in homes and offices and of smartphones in the pockets of individuals has facilitated the transmission of data tremendously. New developments such as network slicing with 5G, Bluetooth Low Energy, LiFi (wireless communication using light), and LoRa (wireless data communication over ranges up to ten kilometers with low power consumption) promise big efficiency improvements in the future. We would also put blockchain technology in this category. Blockchains guarantee the veracity of the data that is being transmitted, adding an important level of trust to transactions that are carried out over networks.

ANALYZING TECHNOLOGIES

The rapid decrease in costs of computing, data storage, and transmission have made cloud-based solutions feasible and accessible to all organizations, regardless of their size. Every firm now has access to computing infrastructure that provides vast storage and tremendous computing power. Partly driven by these technical advances, remarkable progress has been made with respect to analyzing data via machine-learning and deep-learning algorithms. Future improvements in processing power using quantum computing will speed up this development even more.

REACTING TECHNOLOGIES

A whole range of technological advances are continuously pushing down the costs of reacting to customer requests. For instance, improvements in artificial intelligence are allowing automated responses at a vast scale that are becoming more and more personalized. Augmented reality can also be a very effective way of responding to a request by providing the user with rich information. Improvements in 3-D printing and advanced robotics decrease the cost of production at a low scale, and advances in autonomous vehicles and drones are reducing the costs of moving products to customers.

If you get stuck in a particular cell, it can be very helpful to use a classification tree to brainstorm and broaden the set of possible technological solutions for a particular subfunction. Recall from the description in chapter 9 that a classification tree starts with the "job to be done" on the left side and progressively refines the solution space.

After you have filled out worksheet 10-4, you need to decide which particular technological solution to use for each subfunction. To systematize your decision making, it can be helpful to create a selection table, also described in chapter 9. The selection table contains all possible technologies you are considering for a subfunction and your assessments of each technology along various selection attributes, including convenience, reliability, and cost.

Step 5: Identify New Technological Solutions and How Those Might Enable Further Innovations in Your Connected Strategy Not Identified So Far

A big challenge for managers is to keep up with new technological developments. We have found it very helpful to keep track of new technologies by asking what subfunctions a new technology can facilitate.

Possible technological solutions for each subfunction

	Recognize	Request			Respond			Repeat	Connection architecture	Revenue model
	Become aware of the need	Search and decide on option	Order	Pay	Receive	Experience	After sale	Learn and improve	Connect parties in ecosystem	Monetize customer relationship
Sense										
Transmit										
Analyze										
React										

New technological solutions that would enable a new connected strategy

	Recognize	Request			Respond			Repeat	Connection architecture	Revenue model
	Become aware of the need	Search and decide on option	Order	Pay	Receive	Experience	After sale	Learn and improve	Connect parties in ecosystem	Monetize customer relationship
Sense										
Transmit										
Analyze										
React										

Thus, as you discover or read about new technological advances, place them into worksheet 10-5 or use the matrix you have filled out in worksheet 10-4.

Cross-checking this against the "jobs to be done" you outlined earlier can give you insights into possible new ways of implementing your connected strategy: when a new technology emerges, it might enable a new connected relationship or a new connection architecture.

This type of bottom-up innovation is oftentimes referred to as technology push. Rather than thinking about the needs of the customer, you start with a given technology that substantially improves the execution of a function. Then ask yourself, "How might we take advantage of technology X?" For example, you might ask, "How can our business benefit from advances in natural language processing or in augmented reality?" Place the answers to these questions alongside the new technologies in worksheet 10-5.

Workshop Summary

You have reached the end of the last workshop!

Now is a good time to recap how these workshops connect to each other. In workshop 1, you sketched out the efficiency frontier for your industry and located your firm and your competitors relative to the frontier. Using a connected strategy, your goal is to push out this frontier and break the existing trade-off between willingness-to-pay and fulfillment costs.

In workshop 2, we focused on creating a connected customer relationship that will allow you to increase the willingness-to-pay of your customers. To build such a relationship requires a deep understanding of the entire customer journey so that you can refine your ability to recognize your customers' needs, translate those needs into an actionable request for a desired solution, and respond in a timely and frictionless manner. Lastly, if you can repeat this interaction many times, you might be able to climb up the hierarchy of needs of your customers, allowing you to become a trusted partner.

In workshop 3, we focused on how you can create a connected delivery model that allows you to create connected customer relationships at low fulfillment costs. To this end, we asked you to think about different connection architectures, revenue models, and technology infrastructure solutions found in other industries. We hope that this trilogy of workshops has helped you apply the concepts of connected strategy to your own organization.

We did our best to bring to life the interactive style of knowledge transfer that we love so much in our work with students and executive education audiences, but in this case using a very old and very unconnected technology—a book. We already pointed you to our website (connected-strategy.com) as one form of increased connectivity. The site has a substantial amount of curated content and also allows you to post questions and make suggestions for future content updates and new podcasts.

Did we succeed in guiding you through the workshops, and have you thus developed your own connected strategy? If you already worked through some (or, ideally, all) of the three workshops in this book, congratulations—you are done with the homework assignments. Give yourself an A+. We are proud of you.

If you read through this book but skipped the workshops ("I will get to this later"), maybe feeling intimidated by the many questions and exhibits or simply being too busy to do more than just read, please read on for just one more page.

In our view, simply reading about connected strategies without taking any actions is just like sitting next to the pool wondering whether you should get wet. If this adequately captures your position, allow us to give you a final nudge (not push), hoping that you jump in.

The hardest part in taking any journey (including strategic planning processes) is taking the first step. In this spirit, let us propose a set of alternative mini workshops with the request that you do one.

This should not take more than ten (!) minutes. So we are not asking you to jump into the water, just to dip your toe in. Specifically, pick one of the following tasks and then decide for yourself whether you want to do more:

- Share the example of Disney's MagicBand with a colleague on your management team and discuss what you can learn from this.

- As a customer, sign up with a firm that provides a connected relationship you have not experienced before.

- Ask one of your customers how she experiences the episodes during which she connects to your firm, and then ask yourself how your firm connects with this customer.

- Discuss with one of your employees how efficiency could be improved through better connectivity.

- Ask somebody on your R&D or technology team about what technologies they are currently working on and how that might improve the connectivity to your customers.

Take a shot at one of them and see where this leads you . . .

Seizing the Connected Strategy Potential

After a great family day in the Disney theme park, you head back to your hotel. Just as you leave, you meet him again. Captain Jack Sparrow calls out to your six-year-old by name and waves goodbye, capping a truly magical day. The power of these connected relationships is something to ponder. Ordering your food without worrying about payment logistics, following a customized itinerary through the park, and anticipating the great photo collection you can browse on your flight home (thanks to the automated documentation of your visit without your ever having to pull out a camera)—all of these have got you smiling and thinking about how you could use connected strategies to create competitive advantage for your business. Jack Sparrow never met your child before and was relying on your MagicBand to reveal your family's identity. But that doesn't bother you, nor do you mind Disney's healthy profit margins thanks to the remarkable efficiency of the park's operation. The magic of the connected strategy has your attention.

Now imagine Jack Sparrow turns to you and says in his best gravelly pirate voice, "Thanks for your visit, matey. I found access to your financial accounts and I saw that your asset allocation seemed to be underdiversified. So I sold some of your bonds and bought gold for you. Arr, what else would you expect a pirate to buy?" This entirely fictitious example is a textbook application of our connected strategy framework. Jack recognized a potential need, found a solution, and acted promptly. In fact, he created an automatic execution customer experience for you. While some might be delighted, chances are you are not amused.

In this closing section, we want to reiterate a few critical points. Connected strategy does not mean automating every possible transaction for every customer. Automatic execution has its place and will become feasible across ever-broader domains, but for many transactions what customers want is help in making better decisions, not someone (or something) else who makes decisions for them. The key to success for connected strategies lies in understanding the connectivity preferences of your customers. Some customers will enjoy receiving an automatically created photo album of their trip; others will find it an invasion of privacy. Some like to be coached in their behavior and derive value from being nudged; others find it overbearing. And no customer likes his or her data to be misused. In the excitement around technological advances that enable new connected strategies, it is important to resist the temptations to create connectivity just because we can and to monetize data indiscriminately just because no one is stopping us yet. Put your customer first and remember that not all your customers are the same. Building trust is at the heart of a connected strategy, and that trust can be easily lost. Trouble can ensue unless you show that the data your customer gave you creates value for him or her and is not used in undesired ways.

Another misconception that we have hopefully removed is that a connected strategy is primarily about technology. Clearly, technology plays an important role, and quite often new technology is what allows new connected strategies to arise. But as we have described in detail, the connected strategy is fundamentally a business model innovation. Not only will you need to adopt new technologies, but you may also have to change whom you interact with, how you charge for products and services, and how you structure your company internally. To build a connected strategy often requires restructuring your company to ensure the frictionless flow of information within your organization and between you and your customer.

With the ability to increase the customers' willingness-to-pay while reducing fulfillment costs, connected strategies have proved to be truly disruptive in several industries. For you, this is both an opportunity and a threat. We hope that our connected strategy framework and our workshop chapters will help you in creating your own connected strategy. At the same time, we hope that these tools will help you to look at changes in your industry from a new perspective, separating technological hype from true strategic challenges.

There is no doubt that the dramatic increase in connectivity will continue. You may or may not be thinking about connected strategies in your industry, but it's certain that others are. We have only seen the very beginning of connected strategies!

SOURCES

Prologue

For more details on how Disney introduced the **MagicBand**, listen to the podcast on our book website, connected-strategy.com.

For more information on the **MagicBand**, see Austin Carr, "The Messy Business of Reinventing Happiness," *Fast Company*, April 15, 2015, https://www.fastcompany.com/3044283/the-messy-business-of-reinventing-happiness; and Christian Terwiesch and Nicolaj Siggelkow, "When Fun Goes Digital: Creating the Theme Park of the Future," *Knowledge@Wharton*, April 4, 2018, http://knowledge.wharton.upenn.edu/article/future-theme-park-innovation/.

For a thorough discussion on how **interconnected, smart devices** can affect competition and strategy, see Michael E. Porter and James E. Heppelman, "How Smart, Connected Products Are Transforming Competition," *Harvard Business Review*, November 2014, 64–88.

Chapter 1

For very insightful early work on **strategy in the context of the internet**, see Michael E. Porter, "Strategy and the Internet," *Harvard Business Review*, March 2001, 62–78; Raphael Amit and Christoph Zott, "Value Creation in E-business," *Strategic Management Journal* 22, no. 6–7 (2001): 493–520; and Adrian J. Slywotzky, Karl Weber, and David J. Morrison, *How Digital Is Your Business?* (New York: Crown Business, 2000).

For more detail on **business models and business model design**, see the following works by Raphael Amit and Christoph Zott: "Business Model Design and the Performance of Entrepreneurial Firms," *Organization Science* 18, no. 2 (2007): 181–199; "The Fit between Product Market Strategy and Business Model: Implications for Firm Performance," *Strategic Management Journal* 29, no. 1 (2008): 1–26; "Business Model Design: An Activity System Perspective," *Long Range Planning* 43, nos. 2–3 (2010): 216–226.

For a discussion of the **willingness-to-pay concept** and foundational work on the concept of "value," see Adam M. Brandenburger and Harborne W. Stuart, Jr., "Value-Based Business Strategy," *Journal of Economics and Management Strategy* 5, no. 1 (1996): 5–24; and Adam M. Brandenburger and Barry J. Nalebuff, *Coopetition* (New York: Doubleday, 1996).

The **customer journey** has its intellectual foundation in the work by Ian McMillan and Rita McGrath on the consumption chain. See Ian MacMillan and Rita Gunther McGrath, "Discovering New Points of Differentiation," *Harvard Business Review*, July–August 1997, 133–138, 143–145; and Rita Gunther McGrath and Ian MacMillan, *The Entrepreneurial Mindset; Strategies for Continuously Creating Opportunity in an Age of Uncertainty* (Boston: Harvard Business School Press, 2000).

For a wide-ranging discussion of **new technologies** and their impact on industries, firms, and society, see Erik Brynjolfsson and Andrew McAfee, *The Second Machine Age: Work, Progress, and Prosperity in a Time of Brilliant Technologies* (New York: Norton, 2016); and Andrew McAfee and Erik Brynjolfsson, *Machine, Platform, Crowd: Harnessing Our Digital Future* (New York: Norton, 2017).

The pioneering work of our colleagues at Penn, in particular Kevin G. Volpp and David A. Asch, provides a number of great ideas for how **patient care** might be redesigned through the introduction of principles of behavioral economics into the domain of medicine. We discuss several examples of their research in other chapters.

The work by Christian Terwiesch, David Asch, and Kevin Volpp looks at problems of introducing **connected health care** in greater detail. See Christian Terwiesch, David Asch, and Kevin Volpp, "Technology and Medicine: Reimagining Provider Visits as the New Tertiary Care," *Annals of Internal Medicine* 167, no. 11 (2017): 814–815.

For a delightful exposition on the origins of the phrase "**on the shoulders of giants**," see Robert K. Merton, *On the Shoulders of Giants: A Shandean Postscript* (New York: Free Press, 1965).

For more on **platform strategies**, see Marshall W. Van Alstyne, Geoffrey G. Parker, and Sangeet Paul Choudary, "Pipelines, Platforms, and the New Rules of Strategy," *Harvard Business Review*, April 2016, 54–60.

For the number of active **Pokemon Go players**, see Craig Smith, "85 Incredible Pokemon Go Statistics and Facts," DMR, May 5, 2018, https://expandedramblings.com/index.php/pokemon-go-statistics/.

For more details on **Rolls-Royce's transition from products to services**, see "Power by the Hour," Rolls-Royce, accessed October 24, 2018, https://www.rolls-royce.com/media/our-stories/discover/2017/totalcare.aspx; and "Rolls-Royce and Microsoft Collaborate to Create New Digital Capabilities," Microsoft, April 20, 2016, https://customers.microsoft.com/en-US/story/rollsroycestory.

Chapter 2

For the size of the **US grocery industry**, see "Supermarket Facts," FMI, accessed June 17, 2018, https://www.fmi.org/our-research/supermarket-facts.

For the size of the **Indian grocery market** and information on BigBasket, see "Why India's Online Grocery Battle Is Heating Up," *Knowledge@Wharton*,

April 26, 2018, http://knowledge.wharton.upenn.edu/article/indias-online-grocery
-battle-heating/?utm_source=kw_newsletter&utm_medium=email&utm
_campaign=2018-05-01.

Gerard Cachon and Christian Terwiesch discuss the **classification of willingness-
to-pay drivers** in further detail in *Operations Management* (New York: McGraw-Hill
Education, 2016).

For more on the concept of the "**efficiency frontier**," see Michael E. Porter, "What
Is Strategy?," *Harvard Business Review*, November–December 1996, 61–78.

For more information on **Blue Apron**, see Sarah Halzack, "Why This Start-up
Wants to Put Vegetables You've Never Heard of on Your Dinner Table," *Washing-
ton Post*, June 15, 2016, https://www.washingtonpost.com/news/wonk/wp/2016/06
/15/why-this-start-up-wants-to-put-vegetables-youve-never-heard-of-on-your
-dinner-table/.

The **efficiency frontier can be shifted** by better matching supply with demand.
This is also the title of the book written by Cachon and Terwiesch: Gerard Cachon
and Christian Terwiesch, *Matching Supply with Demand: An Introduction to Operations Management* (New York: McGraw Hill Education, 2012).

An interview with an **Instacart** executive can be found on our book website,
connected-strategy.com.

For more information on **Alibaba's Hema stores**, see Christine Chou, "Alibaba to
Open 3 Hema Stores in Xian in 2018," Alizila, January 18, 2018, http://www.alizila
.com/alibaba-open-3-hema-stores-xian-2018/.

Pareto dominance also plays an important role in the notion of an efficient frontier in finance. When making financial investments and allocating assets to investment portfolios, investors typically face a risk-reward trade-off. Investments that
have high expected returns, such as investments in venture capital or private eq-
uity, typically also come with more risk (technically speaking, the standard deviation of the returns is larger). According to Nobel laureate Harry Markowitz, the
efficient frontier is characterized by those investment portfolios for which no
other portfolios exists that achieves a higher expected return for the same standard deviation of return (risk). Investors care about both characteristics of their
portfolio, expected returns (that they would like to be high) and risk (that they
would prefer to be low). Some investors might prefer an expected return of
10% and a standard deviation of returns of 4% over one with an expected return
of 6% and a standard deviation of 2%. Some more risk-averse investors might prefer it the other way around. However, no rational investor would prefer an expected return of 6% and a standard deviation of 4% over one with 10% expected
return and a standard deviation of 2%. It is said that the former is Pareto dominated by the latter. Thus, the efficient frontier is the set of investments that are
not Pareto dominated. For the foundational treatise, see H. M. Markowitz, "Portfolio Selection," *Journal of Finance* 7, no. 1 (1952): 77–91.

For times at which **surge pricing** happens for Uber, see Peter Cohen, Robert
Hahn, Jonathan Hall, Steven Levitt, and Robert Metcalfe, "Using Big Data to Es-
timate Consumer Surplus: The Case of Uber" (NBER Working Paper No. 22627,

National Bureau of Economic Research, Cambridge, MA, September 2016), https://doi.org/10.3386/w22627.

The details of **efficiency** and its relationship to capacity utilization are discussed as part of a lean operations framework in the book written by Cachon and Terwiesch: Gerard Cachon and Christian Terwiesch, *Operations Management* (New York: McGraw-Hill Education, 2016).

All data on the **New York City cab operations** is taken from the reports issued by the New York Taxi and Limousine Commission, accessed June 18, 2018, http://www.nyc.gov/html/tlc/html/technology/industry_reports.shtml.

All **Uber data** is taken from a case study written by our colleague Gerard Cachon, "Uber: Charging Up and Down," The Wharton School, 2018, as well as from Jonathan V. Hall and Alan B. Kruger, "An Analysis for the Labor Market for Uber's Driver-Partners in the United States" (working paper, January 2015), http://arks.princeton.edu/ark:/88435/dsp010z708z67d, and Judd Cramer and Alan B. Kruger, "Disruptive Change in the Taxi Business: The Case of Uber" (NBER working paper No. 22083, National Bureau of Economic Research, Cambridge, MA, March 2016), https://www.nber.org/papers/w22083.

An interview with a **HomeAway** executive can be found on our book website, connected-strategy.com.

For more information on **Goodr**, featured in the sidebar, see Ben Paynter, "This App Delivers Leftover Food to the Hungry, Instead of to the Trash," *Fast Company*, May 3, 2018, https://www.fastcompany.com/40562448/this-app-delivers-leftover-food-to-the-hungry-instead-of-the-trash.

For **average vehicle occupancy**, see the National Household Travel Survey by the Federal Highway Administration: "National Household Travel Survey," US Department of Transportation Federal Highway Administration, accessed June 18, 2018, https://nhts.ornl.gov/.

For data on **Match.com**, see "Match.com Information, Statistics, Facts and History," Dating Sites Reviews, last modified May 28, 2018, https://www.datingsitesreviews.com/staticpages/index.php?page=Match-com-Statistics-Facts-History.

The elements of **willingness-to-pay** have also been discussed by Cachon and Terwiesch in *Operations Management*.

For the **Target** story, see Charles Duhigg, "How Companies Learn Your Secrets," *New York Times Magazine*, February 16, 2012, https://www.nytimes.com/2012/02/19/magazine/shopping-habits.html.

For more information on **Cambridge Analytica**'s use of Facebook data, see Matthew Rosenberg, Nicholas Confessore, and Carole Cadwalladr, "How Trump Consultants Exploited the Facebook Data of Millions," *New York Times*, March 17, 2018, https://www.nytimes.com/2018/03/17/us/politics/cambridge-analytica-trump-campaign.html.

For information on **Google apps using location data**, see Ryan Nakashima, "Google Tracks Your Movements, Like It or Not," AP News, August 13, 2018, https://apnews.com/828aefab64d4411bac257a07c1af0ecb.

For more on the topic of **whether Uber drivers should be considered employees**, see Daniel Wiessner, "US Judge Says Uber Drivers Are Not Company's Employees," Reuters Business News, April 12, 2018, https://www.reuters.com/article/us -uber-lawsuit/u-s-judge-says-uber-drivers-are-not-companys-employees -idUSKBN1HJ31I.

For the **clash between Airbnb and local cities** concerning regulation, see Scott Zamost, Hannah Kliot, Morgan Brennan, Samantha Kummerer, and Lora Kolodny, "Unwelcome Guests: Airbnb, Cities Battle over Illegal Short-Term Rentals," CNBC News, May 24, 2018, https://www.cnbc.com/2018/05/23/unwelcome -guests-airbnb-cities-battle-over-illegal-short-term-rentals.html.

For the use of **Amazon Echo** recordings in a murder case, see Eliott C. McLaughlin, "Suspect OKs Amazon to Hand Over Echo Recordings in Murder Case," CNN, April 26, 2017, https://www.cnn.com/2017/03/07/tech/amazon-echo-alexa-bentonville -arkansas-murder-case/index.html.

Chapter 3

For the argument that **power tools compete with neckties**, see Michael E. Porter, "The Five Competitive Forces That Shape Strategy," *Harvard Business Review,* January 2008, 79–93.

Chapter 4

For more on the expanded role of "**search**"—e.g., in creating curated offering— see Stefan Weitz, *Search* (Brookline, MA: Bibliomotion, 2014).

For more details on the examples in the sidebar on **true personalization**, see "Cosmetics Industry in the U.S.—Statistics & Facts," Statista, accessed June 18, 2018, https://www.statista.com/topics/1008/cosmetics-industry/. Also see "Shiseido Americas Announces bareMinerals® First Brand to Launch Customized by MATCHCo Technology with the Introduction of the MADE-2-FIT App for iPhone®," Cision PR Newswire, June 20, 2017, https://www.prnewswire.com /news-releases/shiseido-americas-announces-bareminerals-first-brand-to-launch -customized-by-matchco-technology-with-the-introduction-of-the-made-2-fit-app -for-iphone-300476833.html. For more information, see Rina Raphael, "Is Customization the Future of the Beauty Industry?," *Fast Company*, October 14, 2016, https://www.fastcompany.com/3064239/is-customization-the-future-of-the-beauty -industry. For information on **3-D printing of drugs**, see Benedict, "$5K Vitae Industries AutoCompounder Can 3D Print Personalized Drugs in 10 Minutes," 3ders.org, December 15, 2017, https://www.3ders.org/articles/20171215-vitae -industries-autocompounder-can-3d-print-personalized-drugs-in-10-minutes .html. See also "The Future of 3D Printing Drugs in Pharmacies Is Closer Than You Think," Medical Futurist, May 4, 2017, http://medicalfuturist.com/future-3d -printing-drugs-pharmacies-closer-think/.

For more detail on the examples in the sidebar on **coach behavior wearable sensors**, see "Discover Two New Wearable Technologies," La Roche-Posay, accessed

June 18, 2018, https://www.laroche-posay.us/wearable-tech.html; and Michael Sawh, "The Best Smart Clothing: From Biometric Shirts to Contactless Payment Jackets," Wareable, April 16, 2018, https://www.wareable.com/smart-clothing/best -smart-clothing.

For more details on the examples in the sidebar on **automated execution with video games**, see Dylan Matthews, "Humans Have Spent More Time Watching Gangnam Style Than Writing All of Wikipedia," Vox, June 7, 2014, https://www .vox.com/2014/6/7/5786480/humans-have-spent-more-time-watching-gangnam -style-than-writing-all. For information on how long individual players spend playing *World of Warcraft*, see Pin-Yun Tarng, Kuan-Ta Chen, and Polly Huang, "An Analysis of WoW Players' Game Hours," *NetGames 2008: Proceedings of the 7th ACM SIGCOMM Workshop on Network and System Support for Games* (2008): 47–52, http://www.iis.sinica.edu.tw/~swc/pub/wow_player_game_hours.html. For more information on the **gaming industry**, see Robert Lee Hotz, "When Gaming Is Good for You," *Wall Street Journal*, March 13, 2012. For a classification of **player types**, see Richard Bartle, "Hearts, Clubs, Diamonds, Spades: Players Who Suit MUDs," Mud.co.uk, August 28, 1996, http://mud.co.uk/richard/hcds.htm. For the concept of "**flow**," see Mihaly Csikszentmihalyi, *Flow: The Psychology of Optimal Experience* (New York: HarperCollins, 1991).

In medicine, the term *automatic hovering* was introduced in a *New England Journal of Medicine* article by David A. Asch, Ralph W. Muller, and Kevin G. Volpp. See David A. Asch, Ralph W. Muller, and Kevin G. Volpp, "Automated Hovering in Health Care—Watching Over the 5000 Hours," *New England Journal of Medicine* 367 (2012): 1–3.

Chapter 5

For the **number of teachers** in the United States, see "Fast Facts," National Center for Education Statistics, accessed June 18, 2018, https://nces.ed.gov/fastfacts/display .asp?id=372.

For more on the **Khan Academy**, see "About," Khan Academy, accessed June 18, 2018, https://www.khanacademy.org/about.

An interview with a **Lynda.com** executive can be found on our book website, connected-strategy.com.

The Mack Institute report by Christian Terwiesch and Karl T. Ulrich discusses the disruptive potential of **online teaching**. See Christian Terwiesch and Karl T. Ulrich, "Will Video Kill the Classroom Star? The Threat and Opportunity of Massively Open Online Courses for Full-Time MBA Programs," Mack Institute for Innovation Management, July 16, 2014, http://www.ktulrich.com/uploads/6/1/7 /1/6171812/terwiesch-ulrich-mooc-16jul2014.pdf.

An interview with a **Rosetta Stone** executive can be found on our book website.

For more information on **artificial intelligence and deep learning**, see Robert D. Hof, "Deep Learning: Artificial Intelligence Is Finally Getting Smart," *MIT Technology Review*, 2013, https://www.technologyreview.com/s/513696/deep-learning/.

For an insightful discussion of how **increased connectivity** among products and services requires changes within organizations, see Michael E. Porter and James E. Heppelman, "How Smart, Connected Products Are Transforming Companies," *Harvard Business Review*, October 2015, 97–114.

For the **market share of Amazon**, at least when we were writing this book, see Rani Molla, "Amazon Could Be Responsible for Nearly Half of U.S. E-commerce Sales in 2017," Recode, October 24, 2017, https://www.recode.net/2017/10/24 /16534100/amazon-market-share-ebay-walmart-apple-ecommerce-sales-2017.

For more information on the costs of **nonadherence for asthma medication**, mentioned in the sidebar, see Aurel O. Iuga and Maura J. McGuire, "Adherence and Health Care Costs," *Risk Management and Healthcare Policy* 7 (2014): 35–44; and *Adherence to Long-Term Therapies: Evidence for Action* (Geneva: World Health Organization, 2013), http://www.who.int/chp/knowledge/publications/adherence _full_report.pdf.

We found the material from the Stanford d.School very helpful in articulating the framework of the **how-why ladder**. See "Welcome," Stanford d.School, accessed June 18, 2018, https://dschool.stanford.edu/.

For the study concerning **adherence to medication after cardiac treatment**, see Kevin G. Volpp et al., "Effect of Electronic Reminders, Financial Incentives, and Social Support on Outcomes after Myocardial Infarction," *JAMA Internal Medicine* 177, no. 8 (June 2017): 1093–1101.

For more information on **Netflix** and a list of Netflix genres, see Rani Molla, "Netflix Now Has Nearly 118 Million Streaming Subscribers Globally," Recode, January 22, 2018, https://www.recode.net/2018/1/22/16920150/netflix-q4-2017 -earnings-subscribers; and "Complete Searchable List of Netflix Genres with Links," Finder, last modified June 6, 2018, https://www.finder.com/netflix/genre -list.

For the **Organisation for Economic Co-operation and Development guidelines**, see *The OECD Privacy Framework* (OECD, 2013), http://www.oecd.org/sti/ieconomy /oecd_privacy_framework.pdf. Also see Anna E. Shimanek, "Do You Want Milk with Those Cookies? Complying with Safe Harbor Privacy Principles," *Journal of Corporation Law* 26, no. 2 (2001): 455, 462–463.

For the **EU guidelines**, see "GDPR Key Changes," EUGDPR, accessed November 24, 2018, https://eugdpr.org/the-regulation/.

Chapter 7

For data on **car2go**, see "Get In and Drive Off: Free-Floating Carsharing with car2go," Daimler, accessed June 18, 2018, https://www.daimler.com/products /services/mobility-services/car2go/. For information on the merger with **BMW's ReachNow**, see Nat Levy, "Car2go and ReachNow Car-Sharing Services to Merge in Deal between Auto Giants Daimler, BMW," Geekwire, March 28, 2018, https:// www.geekwire.com/2018/car2go-reachnow-car-sharing-services-merge-deal-auto -giants-daimler-bmw/.

For more information on **Carnival's medallion**, see Brooks Barnes, "Coming to Carnival Cruises: A Wearable Medallion That Records Your Every Whim," *New York Times*, January 4, 2017, https://www.nytimes.com/2017/01/04/business/media/coming -to-carnival-cruises-a-wearable-medallion-that-records-your-every-whim.html.

For more details on the examples in the sidebar on **e-scooter and battery sharing**, see Bérénice Magistretti, "Gogoro Raises $300 Million for Its Battery-Swapping Technology," VentureBeat, September 19, 2017, https://venturebeat.com/2017/09 /19/gogoro-raises-300-million-for-its-battery-swapping-technology/; and Karen Hao, "The Future of Transportation May Be about Sharing Batteries, Not Vehicles," Quartz Media, September 25, 2017, https://qz.com/1084282/the-future-of -transportation-may-be-about-sharing-batteries-not-vehicles/. Bike sharing and e-scooter sharing are also discussed in one of the podcasts on our book website, connected-strategy.com.

For more details on the sidebar on the **Pandora of the art world**, see Molly Schuetz, "New York's Artsy Is Making It Even Easier to Buy Art Online," Bloomberg, March 27, 2018, https://www.bloomberg.com/news/articles/2018-03-27/new -york-s-artsy-is-making-it-even-easier-to-buy-art-online. For more details on the art classification scheme that **Artsy** has developed, see "The Art Genome Project," Artsy, accessed June 18, 2018, https://www.artsy.net/categories; and Shahan Mufti, "Artsy's 'Genome' Predicts What Paintings You Will Like," *Wired*, November 23, 2011, https://www.wired.com/2011/11/mf_artsy/all/1/.

A thorough discussion of direct-to-consumer companies, and the observation that **customer acquisition costs are the new rent**, see Tom Foster, "Over 400 Startups Are Trying to Become the Next Warby Parker. Inside the Wild Race to Overthrow Every Consumer Category," *Inc.*, May 2018, https://www.inc.com/magazine/201805 /tom-foster/direct-consumer-brands-middleman-warby-parker.html.

For data on **Kickstarter**, see "Stats," Kickstarter, accessed November 23, 2018, https://www.kickstarter.com/help/stats.

For data on **DonorsChoose.org**, see "Impact," DonorsChoose.org, accessed November 23, 2018, https://www.donorschoose.org/about/impact.html.

For more information on **Crisis Text Line**, see Alice Gregory, "R U There? A New Counselling Service Harnesses the Power of the Text Message," *New Yorker*, February 9, 2015, https://www.newyorker.com/magazine/2015/02/09/r-u. For access to **data collected by Crisis Text Line**, see "Crisis Trends," Crisis Text Line, accessed November 23, 2018, https://crisistrends.org/.

Chapter 8

For average spending on **dental care**, see Health Policy Institute, "U.S. Dental Expenditures: 2017 Update," American Dental Association, accessed June 18, 2018, https://www.ada.org/~/media/ADA/Science%20and%20Research/HPI/Files /HPIBrief_1217_1.pdf?la=en.

For more information on **Abilify**, see "FDA Approves Pill with Sensor That Digitally Tracks If Patients Have Ingested Their Medication," US Food and Drug Administration, November 13, 2017, https://www.fda.gov/NewsEvents/Newsroom /PressAnnouncements/ucm584933.htm.

For more information on **Fitbit**, see Stephanie Baum, "Fitbit Plans to Submit Sleep Apnea, AFib Detection Tools for FDA Clearance," MedCity News, February 27, 2018, https://medcitynews.com/2018/02/fitbit-plans-to-submit-sleep-apnea-afib -detection-tools-for-fda-clearance/.

In the context of the **limits of traditional revenue models**, we found the work by Karan Girotra and Serguei Netessine very insightful. In their book on business model innovation, they outline how to create a business model that is more robust in an uncertain environment. The authors discuss how value can be created by overcoming these inefficiencies. See Karan Girotra and Serguei Netessine, *The Risk-Driven Business Model: Four Questions That Will Define Your Company* (Boston: Harvard Business Review Press, 2014).

For the study concerning reimbursement of **retinal photographs**, see David A. Asch, Christian Terwiesch, and Kevin G. Volpp, "How to Reduce Primary Care Doctors' Workloads while Improving Care," *Harvard Business Review*, November 2017.

For more information on **Rolls-Royce**'s revenue model, see "'Power by the Hour': Can Paying Only for Performance Redefine How Products Are Sold and Serviced?," *Knowledge@Wharton*, February 21, 2007, http://knowledge.wharton.upenn .edu/article/power-by-the-hour-can-paying-only-for-performance-redefine-how -products-are-sold-and-serviced/.

The work by Marco Iansiti and Roy Levien further discusses the idea of **moving the focus from the supply chain to the ecosystem**. See Marco Iansiti and Roy Levien, "Strategy as Ecology," *Harvard Business Review*, March 2004, 68–78, 126.

For a great discussion on when and how **freemium** models work, see Vineet Kumar, "Making 'Freemium' Work," *Harvard Business Review*, May 2014, 27–29.

For more detail on **Tencent and WeChat**, see Eveline Chao, "How WeChat Became China's App for Everything," *Fast Company*, January 2, 2017, https://www .fastcompany.com/3065255/china-wechat-tencent-red-envelopes-and-social-money.

For data on **in-app purchases**, see "These 25 Wildly Popular Android Games Are Raking in the Most Cash from In-App Purchases," ZDNet, April 17, 2017, https:// www.zdnet.com/pictures/25-wildly-popular-android-games-raking-in-the-most -cash-from-in-app-purchases/26/.

For the number of **Amazon Prime** members, see Heather Kelly, "Amazon Reveals It Has More Than 100 Million Prime Members," CNN Tech, April 19, 2018, http:// money.cnn.com/2018/04/18/technology/amazon-100-million-prime-members /index.html.

For the prices of **Google AdWords**, see Elisa Gabbert, "The 25 Most Expensive Keywords in AdWords—2017 Edition!," *WordStream Blog*, last updated September 12, 2018, https://www.wordstream.com/blog/ws/2017/06/27/most-expensive -keywords.

Chapter 9

The product development book by Karl T. Ulrich and Steven D. Eppinger explains several approaches to **deconstruction** (the authors talk about decomposition). In its chapter on concept generation, the book provides an excellent guide for how to generate a large number of concepts in response to a given set of needs. See Karl T. Ulrich and Steven D. Eppinger, *Product Design and Development*, 6th ed. (New York: McGraw-Hill Education, 2015).

For more details on the sidebar on **applying the STAR approach to a new schizophrenia drug**, see Kanika Monga and Olivia Myrick, "Digital Pill That 'Talks' to Your Smartphone Approved for First Time," ABC News, November 15, 2017, http://abcnews.go.com/Health/digital-pill-talks-smartphone-approved-time/story ?id=51161456.

In table 9-1, we note that one "job to be done" is to "[a]ssess delight of user with coffee brand using physiological measurements." For work in this direction, see Cipresso Pietro, Serino Silvia, and Riva Giuseppe, "The Pursuit of Happiness Measurement: A Psychometric Model Based on Psychophysiological Correlates," *Scientific World Journal* 2014 (2014): 1–15, http://dx.doi.org/10.1155/2014/139128.

The computer network textbook by Andrew S. Tanenbaum and David J. Wetherall provides a detailed discussion of the **stack framework**, especially the Open Systems Interconnection model. See Andrew S. Tanenbaum and David J. Wetherall, *Computer Networks*, 5th ed. (Upper Saddle River, NJ: Prentice Hall, 2010). See also Norman F. Schneidewind, *Computer, Network, Software, and Hardware Engineering with Applications* (Hoboken, NJ: Wiley-IEEE Press, 2012).

The **classification tree framework** is also inspired by Ulrich and Eppinger, *Product Design and Development*. It provides an effective method for exploring a broad array of design approaches for a given design problem. In his book *Design: Creation of Artifacts in Society* (Philadelphia: University of Pennsylvania, 2011), Karl T. Ulrich further examines the power of exploring a design space with a tree-based approach.

Christian Terwiesch and Karl Ulrich, in *Innovation Tournaments: Creating and Selecting Exceptional Opportunities* (Boston: Harvard Business School Press, 2009), distinguish between two types of innovation techniques, **internal and external search**. External search is about scanning the environment for new ideas, oftentimes benefiting from domain arbitrage. A technology might be well established in one industry but still not be used at all in another industry.

The concept of a **selection table** is discussed in *Product Design and Development* by Ulrich and Eppinger in the context of product development. The underlying approach of comparing a set of decision alternatives along a number of criteria is often referred to as multiattribute decision making and has a long tradition in decision sciences.

For a history of **voice recognition**, see Melanie Pinola, "Speech Recognition through the Decades: How We Ended Up with Siri," *PCWorld*, November 2, 2011, https://www.pcworld.com/article/243060/speech_recognition_through_the _decades_how_we_ended_up_with_siri.html.

In *Innovation Tournaments*, Terwiesch and Ulrich discuss how **innovation** can sometimes be triggered by an unmet need but also can occur when a new solution approach is invented. In their work, the authors define *innovation* as a novel match between solution and need.

For information on **digital twins**, see Michael Grieves, "Digital Twin: Manufacturing Excellence through Virtual Factory Replication," white paper, 2014, http://innovate.fit.edu/plm/documents/doc_mgr/912/1411.0_Digital_Twin_White_Paper_Dr_Grieves.pdf.

For more background on **Heller**'s application, see "Appsolute Efficiency," Siemens, accessed October 24, 2018, https://www.siemens.com/customer-magazine/en/home/industry/manufacturing-industry/heller-appsolute-efficiency.html.

For more information on the sidebar on **delivery via drones**, see "Lifesaving Deliveries by Drone," Zipline, accessed June 19, 2018, http://www.flyzipline.com/.

An interview with a **Volocopter** executive can be found on our book website, connected-strategy.com.

For the origins of the **bracelet monitoring device**, see Matt Allyn, "Spider-Man Created the Electronic Bracelet?!," *Esquire*, May 4, 2007, https://www.esquire.com/news-politics/news/a2164/spiderman022007/.

Chapter 10

For an excellent discussion of the various business uses of **augmented reality** for both sensing and responding, see Michael E. Porter and James E. Heppelmann, "Why Every Organization Needs an Augmented Reality Strategy," *Harvard Business Review*, November–December 2017, 46–57.

For the **impact of blockchain technology on trust**, see Kevin Werbach, *The Blockchain and the New Architecture of Trust* (Cambridge: MIT Press, 2018).

For the possible economy-wide impact of **3-D printing**, see Richard D'Aveni, *The Pan-Industrial Revolution: How New Manufacturing Titans Will Transform the World* (Boston: Houghton Mifflin Harcourt, 2018).

INDEX

Note: Page numbers in italics followed by *f*, *t*, and *w* refer to figures, tables, and worksheets, respectively.

Abilify, 176, 204
accessibility, 52–55
Adherium SmartInhaler, 106
advertising revenue, 190–191
AdWords, 190
Airbnb
    crowd orchestrator architecture,
        13, 38, 163–164, 171, 221
    customer experience, 70
Alexa (Amazon), 196, 209
Alibaba, 28, *29f*
alternatives, identification of,
    206–208, *209t*
Amazon
    Alexa, 196, 209
    AmazonFresh, 24
    Amazon Go stores, 29
    Amazon Web Services, 186
    connection architecture, 13, 154,
        156, 158, 171
    customization by, 102
    drone delivery, 213
    Marketplace, 158, 171
    Prime, 189–190
    respond-to-desire customer
        experience, 6, 70
analyzing technologies, 227
Angie's List, 159, 183
Apple, 12, 99, 115, 205, 210, 211

Art Genome Project, 160
artificial intelligence (AI), 84–85
Artsy, 160
Asch, David, 11
asthma inhalers, repeat loop with,
    106
AstraZeneca, 106
attributes, product, 52–53
augmented reality, 14, 42, 228
automatic execution
    characteristics of, 8, 77–79, *78f*
    information flow in, *81t*, 83
    limitations of, 236
    use cases for, 87–88, *88t*
    with video games, 80
automatic hovering, 11, 83, 87,
    109
Avis Budget Group, 155

Bad Ass Mom Box, 154
BareMinerals Made-2-Fit foundation,
    73
BarkBox, 154
Bell Labs, 210
Betterment, 153–154
BigBasket, 27
BlaBlaCar, 11, 40–41
blockchain technology, 39, 227

Blue Apron
  competitive advantage of, 42–43
  curated offering, 71, 154
  disruptive potential of, 47
  efficiency frontier in, 24–26, *27f*, *29f*
BMW, 151
bottom-up innovation, 209–213
bracelet monitoring devices, 208, 215
  See also Disney MagicBand
brainstorming, 51–52
Brandenburger, Adam M., 11
Brother Refresh program, 79
Brynjolfsson, Erik, 11
business-to-business connected
    strategies, 16
Busy Bee Stationery, 154

cab operations, 32–35, 37
Cambridge Analytica, 46
Car2go, 150–151
Careem, 29
Carnival, 152–153
car-sharing services, 150–151, 155
  See also ride-hailing industry
Chinese grocery market, 28–29
Clash of Clans, 187
classification trees, 206–208, *207f*
coach behavior
  characteristics of, 7–8, 73–76, *75f*
  information flow in, *81t*, 82–83
  use cases for, 87, *88t*
competition, ranking, 55–58
competitive advantage, 42–43
  See also repeated interactions
complementary products, revenues
    from, 167
connected customer relationships
  See customer relationships
connected delivery model
  See delivery model
connected market makers, 157–161,
    158f, 171
connected producers, 148–153, *149f*,
    170–171

connected retailers, 153–157, *153f*, 171
connected strategy, definition of, 15
connected-strategy.com website, 17, 49
connected strategy framework,
    1–4, *3f*
connected strategy matrix
  benefits of, 14, 172
  components of, 168–170, *169f*
  mapping, 218–220, *223w*
  strategy development with, 220–222
  worksheets for, *219w*, *223w*
connection architecture
  connected market makers, 157–161,
      158f, 171
  connected producers, 148–153,
      149f, 170–171
  connected retailers, 153–157,
      153f, 171
  crowd orchestrators, 161–165, *162f*,
      171–172
  definition of, 4
  need for, 147–148
  peer-to-peer network creators,
      165–168, *168f*, 172
  role of, 10–11
  See also connected strategy matrix
consumption utility, 52
cost of ownership, 52–55, *54w*
cost reduction
  competitive advantage and,
      42–43
  with deeper connections, 31–35
  diagnostic questions, 50–51
  dynamic pricing strategies, 35–41
  with new connections, 35–41
  potential for, 46–47, 51–52
  See also efficiency frontier; revenue
      models; willingness-to-pay
Couchsurfing, 41
Coup, 156
Crisis Text Line, 165
crowd orchestrators
  connection architecture for,
      161–165, *162f*, 171–172
  not-for-profit, 164–165

curated offering customer
experience
characteristics of, 7, 70–73, *72f*
information flow in, *81t*, 82
use cases for, 87, *88t*
Curb app, 33
customer experiences
artificial intelligence and, 84–85
automatic execution, 8, 77–80, *78f*,
81t, 83, 87–88, *88t*
brainstorming strategies for,
51–52
coach behavior, 7–8, 73–76, *75f*,
81t, 82–83, 87, *88t*
competitive advantage, 42–43
connections, benefits of, 35–41, *38t*
curated offering, 7, 70–73, *72f, 81t*,
82, 87, *88t*
diagnostic questions for, 50–51
disruptive potential of, 46–47
domains of application, 86–88, *88t*
fulfillment costs, 136–141
Goodr case study, 39
information flow, 79–83, *81t*,
131–133
pain points, 31–35, *34t*, 126–130,
136–141
respond-to-desire, 5–6, 67–70, *69f*,
81t, 82, 86–87, *88t*
summary of, 89–90
unified experiences across
episodes, 98–101
See also customer journey;
efficiency frontier; trust;
willingness-to-pay
customer journey
mapping, 122–125
pain points, 31–35, *34t*, 126–130,
136–141
phases of, 63–67, *66f*
worksheet for, *125w*
See also willingness-to-pay
customer relationships
components of, 1–3, *3f*
definition of, 64

goals of, 122
importance of, 8–9
See also customer experiences;
repeated interactions
customization
customer-level learning, 101–103,
103f, 134–136
levels of, 110–111, *111t*, 113,
118–119
population-level learning, 103–107,
107f
why-how ladder, 107–110, *109f*,
133–134
See also curated offering customer
experience

Daimler, 10, 150–151, 171
da shang, 187–188
data payments, 190–193
data privacy, 44–46, 144
dating platforms, 41
deconstruction
business application of, 225–228
dimensions of, 198–201,
202t–203t
worksheet for, *226w*
deep learning, 84–85
delivery model, 3–4, *3f*
See also connection architecture;
revenue models; technology
infrastructure
diagnostic questions, 50–51
Didi, 29
digital twins, 212
Disney MagicBand
competitive advantage of, 2, 235
connection architecture, 6–7,
100–101, 152–153, 170
as innovation of existing
technology, 208, 215
revenue model, 14
rewards of, 34
Dollar Shave Club, 174
DonorsChoose.org, 164

Dragon Dictate, 210
drone delivery, 213
Dropbox, 186
dynamic pricing, 35–41, 178

Earnest, 149–150
EasyTaxi app, 33
eBay, 164
economies of scale, 114–115
ecosystem, 184–185
education, efficiency frontier in,
    94–98, *95f*
efficiency frontier
  competitive advantage and, 5,
    42–43
  concept of, 22–24, *23f*
  disruptive potential of, 47
  in education, 94–98, *95f*
  in grocery retailing, 24–29
  Pareto dominance and, 30, 241
  in ride-hailing industry, 29, 31
  sketching, 55–59
  worksheet for, *56w*
  See also willingness-to-pay
eHarmony, 41
episodic interactions
  See repeated interactions
e-scooter industry, 156
European Union General Data
    Protection Regulation, 117, 193
Expedia, 158–159, 171
external search, 248

Facebook, 13, 46, 159, 161, 190, 192
fees-at-risk revenue model, 184
financial services
  connected producers, 149–150
  crowd orchestrators, 161–162
  customer needs in, 147–148
  peer-to-peer network creators,
    165–168
fit attributes, 52–53
Fitbit, 83, 176–177

Ford, Henry, 91
freemium revenue models, 186–187
friction, 53
  See also cost reduction
fulfillment cost
  definition of, 23
  in education, 95, *95f*
  in meal-kit delivery services, 25–26
  reduction of, 136–141
  See also efficiency frontier

gaming industry, 80, 187
General Data Protection Regulation,
    117, 193
General Motors, 151
GetTaxi app, 33
Gillette, 174
Gobble, 154
Gogoro, 156
Goodr, 39
Google
  AdWords, 190
  connection architecture,
    152, 159, 161
  Nest thermostat, 196
  revenue models, 186, 188, 190
  voice recognition technology,
    210–211
Grab, 29
Green Chef, 154
grocery retailing
  efficiency frontier in, 22–24, *23f*
  willingness-to-pay, 27–29
  See also meal-kit delivery services

haggling, 178
health care space, revenue models
    for
  See revenue models
Heller, 212
HelloFresh, 24, 43, 154
Hema app (Alibaba), 28, *29f*
hidden Markov chains, 210

hierarchy of needs, 108, 112–114, 142, 144, 152, 189, 232
HomeAway, 38
Homeplus app (Tesco), 28, *29f*
hovering, automatic, 11, 83, 87, 109
HP Instant Ink program, 77, 79

IBM, 152, 186, 210
in-app purchases, 187
Indian grocery market, 27
information, limitations in, 178, 180
information flow
  dimensions of, 79–83, *81t*
  identifying and documenting, 131–133
  worksheet for, *132w*
infrastructure-as-a-service revenue model, 186
innovation, 209–213, 236–237, 249
Instacart, 26–27, 171
Instagram, 192
Instant Ink program (HP), 77, 79
Internet of Things, xi, 10, 77, 212
iRobot Roomba, 196
IronPlanet, 159, 160

JD, 29
Jobs, Steve, 205, 208

Kayak, 38–39
Khan, Salman, 92, 95–96
Khan Academy, 92
Kickstarter, 161–162, 171

learning analytics
  customer-level learning, 101–103, *103f*, 134–136
  population-level learning, 103–107, *107f*
  repeat dimension and, 110–115, *112f*

LendingTree, 157
LinkedIn, 167, 186–187, 190
loan providers
  See financial services
L'Oréal, 76
Love, Jack, 215
Lyft, 5, 10, 31, 70, 115
Lynda.com, 93, 102, 107

Mack Institute for Innovation Management, 10, 262–263
MagicBand
  See Disney MagicBand
market makers, connected, 157–161, *158f*, 171
Markowitz, Harry, 241
Match.com, 41, 167
McAfee, Andrew, 11
McGrath, Rita Gunther, 11
McGraw-Hill, 2, 7–9, 102
McMillan, Ian C., 11
meal-kit delivery services
  curated offering, 71
  efficiency frontier in, 24–29, *27f, 29f*
membership revenues, 166–167
micropayment revenue models, 187–188
Microsoft, 186, 210
Mint, 191
MyTaxi app, 33

Nadi X pants, 76
Nalebuff, Barry J., 11
needs, recognition of
  See recognition of needs
Nest thermostat, 196
Netflix, 71, 82, 113, 155, 171
Netsuite, 186
network effects, 115
neural networks, 85
New York City cab operations, 32–35, 37

Niantic, 14
Nike, 7, 167, 170, 186
Nintendo, 14
not-for-profit crowd orchestrators, 164–165

Ola, 29
OnDeck, 149–150
OpenTable, 38, 159
Organisation for Economic Co-operation and Development, 117
ownership, cost of, 52–55

P2P
   See peer-to-peer (P2P) network creators
pain points
   identification of, 126–130, 136–141
   in ride-hailing industry, 31–35
   worksheets for, *130w*, *139w*
Pareto dominance, 30, 241
PatientsLikeMe, 41
pay-as-you-go revenue model, 185–188
pay-for-data revenue model, 192–193
pay-for-performance revenue model, 183–184
pay-with-data revenue model, 190–193
PeachDish, 154
peer-to-peer (P2P) network creators, 165–168, *168f*, 172
performance, pricing contingent on, 183–184
performance attributes, 52
personalization
   See customization
PillsyCap, 176
platform-as-a-service revenue model, 186
platform strategies, 12–13
population-level learning, 103–107, *107f*

Porter, Michael E., 11, 57
posted prices, 178
power purchase agreements (PPAs), 184
Priceline, 158–159, 171
pricing
   dynamic, 178
   posted prices, 178
   surge, 35–41
   See also revenue models
privacy concerns, 44–46, 116–118, 144
producers, connected, 148–153, *149f*, 170–171
product handoff function, 208
Progressive Insurance, 151–152
Prosper, 161
Purple Carrot, 154

QQ Show, 187

reacting technologies, 228
recognition of needs
   in repeated interactions, 98–101
   role of, 3, 64–67, *66t*, 199
   STAR approach to, 200–201, *203t–204t*
   why-how ladder, 107–110, *109f*, 133–134
referral fees, 191
reinvestment into relationships, 189–190
Rent the Runway, 155
repeated interactions
   competitive advantage and, 110–115, *112f*
   customer-level learning in, 101–103, *103f*, 134–136
   data-protection policies, 116–118
   examples of, 91–94
   improving customer experiences with, 141–144

population-level learning in,
    103–107, *107f*
recognition of needs in, 107–110
role in connected strategy
    framework, 3–4, 199
STAR approach to, 200–201,
    202t–203t
summary of, 118–119
unified customer experiences
    across, 98–101
worksheet for, *143w*
See also trust
requests
    role in connected strategy
        framework, 3, 64–67, *66f*, 199
    STAR approach to, 200–201,
        202t–203t
    strengthening through repeated
        interactions, 101–103
respond-to-desire customer
    experience
    characteristics of, 6–7, 67–70, *69f*
    information flow in, *81t*, 82
    use cases for, 86–87, *88t*
response
    role in connected strategy
        framework, 3, 64–67, *66f*, 199
    STAR approach to, 200–201,
        202t–203t
    strengthening through repeated
        interactions, 103–107, *107f*
    See also respond-to-desire
        customer experience
retailers, connected, 153–157, *153f*,
    171
retinal photographs, insurance
    reimbursement of, 181–182
revenue models
    complementary products, 167
    creation of, 222, 224
    definition of, 174
    ecosystem for, 184–185
    examples of, 173–177
    fees for information, 167
    limitations of, 179–181

pay-as-you-go, 185–188
pay-for-data, 192–193
pay-for-performance, 183–184
pay-with-data, 190–193
reinvestment into relationship,
    189–190
role in connected strategy
    framework, 4, 14
summary of, 193–194
traditional, 178–179
transaction or membership
    revenues, 166–167
value creation, 181–183
See also cost reduction
ride-hailing industry
    competitive advantage in, 42–43
    cost reduction in, 35–41
    customer experience in, 5
    disruptive potential of, 47
    efficiency frontier in, 29, 31
    network effect, 115
    pain points in, 31–35, *34t*
    surge pricing in, 35–41
Robert Bosch, 156
Rolls-Royce, 16, 183–184
Roomba, 196

Salesforce, 152, 186
search, external, 248
selection tables, 198, 248
sensing technologies, 225, 227
Sensoria, 76
sensors, wearable, 76
Shiseido, 73
Siri, 210
SmartInhaler, 106
software-as-a-service revenue model,
    186
South Korean grocery market,
    27–28
Square, 104–105
stack framework, 201, 205–206
STAR (sense-transmit-analyze-react)
    approach, 200–201, *202t–203t*

Stuart, Harborne W., Jr., 11
Sun Basket, 154
supply chain, 184
surge pricing, 35–41
Sweeten, 159

Target, 45–46
technology infrastructure
  classification trees for, 206–208,
    207f
  deconstruction of, 198–201,
    202t–203t, 225–228
  design alternatives for, 206–208,
    209t
  identification of, 228–231
  importance of, 14–15
  innovation in, 209–213, 236–237
  role in connected strategy
    framework, 4
  sample scenarios for, 195–198
  stack framework, 201, 205–206
  visionary thinking for, 213–215
  worksheet for, 229w–230w
Tencent, 187–188
Tesco, 27–28, 29f
textbooks, smart, 2, 7–9, 34, 92–93,
  102
three-dimensional printing
  technology, 73, 228
tipping, virtual, 187–188
traditional revenue models,
  178–179
training neural networks, 85
transactional friction, 179, 180
transaction costs, 53
  See also cost reduction
transaction revenues, 166–167
TransferWise, 166
transmitting technologies, 227
TripAdvisor, 190
true personalization, 73
trust
  data-protection policies, 116–118,
    144

importance of, 44–46, 236
revenue models and, 178–180
why-how ladder, 107–110, 109f,
  133–134

Uber
  competitive advantage, 5, 32, 43
  connection architecture, 10, 12,
    170–171
  efficiency frontier, 29
  respond-to-desire customer
    experience, 70
  See also ride-hailing industry
Uberization, 170

Vacation Rental by Owners
  (VRBO), 38
value creation, 181–183
Venmo, 165–166
voice recognition, 209–211
Volkswagen, 151
Volocopter, 213
Volpp, Kevin, 11

Wallaby Financial, 157–158
Wawa, 208
Wealthfront, 153–154
wearable sensors, 76
Wearable X, 76
WeChat, 188
Weinman, Lynda, 93
why-how ladder, 107–110, 109f,
  133–134
willingness-to-pay
  definition of, 22
  drivers of, 52–55
  in education, 95, 95f
  identifying drivers of, 126–130
  in meal-kit delivery services, 25
  in ride-hailing industry, 31–35
  worksheets for, 54w, 130w
  See also efficiency frontier

worksheets
  connected strategy matrix, *219w*,
    223w
  customer journey, *125w*
  deconstruction, *226w*
  efficiency frontier, *56w*
  information flow, *132w*
  pain points, *130w*, *139w*
  repeated customer experiences,
    143w
  technological solutions,
    229w–230w
  why-how ladder, *134w*
  willingness-to-pay drivers, *54w*,
    130w

World of Warcraft, 80
Wozniak, Steve, 205, 208

Xerox PARC, 208

Yandex, 29
YouTube, 167

Zalando, 105, 155
Zipcar, 11, 155–156
Zipline, 213
ZocDoc, 38

ABOUT THE AUTHORS AND ACKNOWLEDGMENTS

Both of us grew up in Germany but pursued (for Germans) uncommon education paths. After receiving a diplom in business information technology at the University of Mannheim, Christian studied at INSEAD in France to receive his PhD in management. Similarly, after completing German high school, Nicolaj went to Stanford University as an undergraduate, receiving a BA in economics, and then on to Harvard, studying under Michael Porter, to receive a PhD in business economics. We both joined the Wharton School of the University of Pennsylvania in 1998 as faculty members, Christian in the Operations, Information and Decisions Department and Nicolaj in the Management Department. Over time, we worked ourselves up the ranks and both now hold endowed chairs. Christian is the Andrew M. Heller Professor and also holds a faculty appointment in Penn's Perelman School of Medicine. Nicolaj is the David M. Knott Professor and a former department chair of Wharton's Management Department.

Our research has appeared in many of the leading academic journals in our fields—Christian focusing on operations management and innovation management, and Nicolaj on strategy and organizational design—including *Management Science*, the *Strategic Management Journal*, *Administrative Science Quarterly*, and the *New England Journal of Medicine*. We are also members of the editorial boards of key academic journals. Christian is the coauthor of *Matching Supply with Demand*, a widely used textbook in operations management, and of *Innovation*

Tournaments, a guide for creating and selecting exceptional opportunities within organizations.

Both of us very much enjoy teaching in our MBA and executive MBA programs. Together, we have won more than fifty teaching awards in the Wharton classroom. One key perk of working at a top business school is the opportunity to present our work in front of many executive audiences. Nicolaj serves as academic director at Wharton Executive Education for two open-enrollment courses on strategy (Creating and Implementing Strategy for Competitive Advantage and Effective Execution of Organizational Strategy), while Christian is academic director for a program on innovation management (Mastering Innovation: From Idea to Value Creation). We have researched with and consulted for more than one hundred different organizations, from small startups to *Fortune* 500 companies.

To achieve broader reach in our teaching efforts, both of us have created online courses as well. Christian was the first to launch a massive open online course (MOOC) in business on Coursera. By now, close to one million students have enrolled in his Introduction to Operations Management course, making this one of the largest online courses. Nicolaj is offering an online course titled Business Strategy from Wharton: Competitive Advantage and the course Strategic Management: Competitive and Corporate Strategy, which is part of Wharton's online certificate on leadership and management.

Another way in which we have attempted to broaden our reach and have the opportunity to learn is by hosting radio shows on Wharton's Sirius XM channel. Christian hosts the show *Work of Tomorrow*, highlighting how technological advances affect the daily operations of many firms, while Nicolaj cohosts *Mastering Innovation*, which addresses how organizations foster innovation that keeps them going strong year after year.

Lastly, both of us are codirectors of the Mack Institute for Innovation Management at Wharton. The Mack Institute's role is to bridge academic research and the world of practice by sponsoring research, conducting conferences, and connecting scholars, business leaders, and students. It was through our work with executive audiences in the classroom and through the Mack Institute that we became inspired to write this book. We observed that firms were connecting to customers in very different ways and were creating new connections between previously unconnected parties in the marketplace. It has been an amazing journey to write this book, allowing us to clarify our thinking and helping others to make more sense of the business world they live in. However, to borrow a phrase from Winston Churchill, we see this book certainly not as the end but only as the end of the beginning of our research into connected strategies. It is an exciting phenomenon that only has just started. Please stay in touch via our website, connected-strategy.com, as our thinking evolves on this subject.

We wouldn't have been able to write this book without the support of many. First of all, we'd like to thank all our MBA and WEMBA students and Executive Education participants at Wharton who over the years have pushed our thinking and forced us to reevaluate how firms create and sustain competitive advantage. The second big thanks goes to Melinda Merino from Harvard Business Review Press, who believed in our project and has given us tremendously helpful feedback throughout the writing and publication process.

We received very valuable research assistance from Matthew Dabaco, Sunil Gottumukkala, Bryan Hong, Jeni Incontro, Ajay Jagannath, Pragna Kolli, Tony Li, Peter Mahon, Josh McLane, Venkat Mendu, Nikhil Nayak, Rohith Ramkumar, Alexandre Teixeira, and Hicham Zahr. Thanks also to four anonymous reviewers who had many helpful suggestions. We are very much indebted to Arik Anderson, Leah Belsky, Carter Cleveland, Justin Dawe, Steven Dominguez, Nick Franklin, John Hass, Josh Hix, Jeff Hurst, Stephan Laster, Sarah Mastrorocco, Eric Merz, Jennifer Neumaier, Florian Reuter, Sonesh Shah, Garth Sutherland, Rambabu Vallabhajosula, Greg Wallace, and Tom Wang for discussing their connected strategies with us. Very valuable editorial help was provided by Angie Basiouny, Greg Bates, Michelle Eckert, and Katherine Fitz-Henry, and Deborah Watson who performed some last-minute heroics. Special thanks are due to Adam Grant, who has graciously shared with us his extensive learnings concerning the publishing experience. Finally, we are very grateful to the Wharton School and the Mack Institute for Innovation Management for financial support.